Bird Pellets

BIRD PELLETS

A COMPLETE PHOTOGRAPHIC GUIDE

ED DREWITT

PELAGIC PUBLISHING

First published in 2024 by
Pelagic Publishing
20–22 Wenlock Road
London N1 7GU, UK
www.pelagicpublishing.com

Bird Pellets: A Complete Photographic Guide

Reprinted 2025

https://doi.org/10.53061/NUXA1756

British Library Cataloguing in Publication Data
A catalogue record for this book is available from the British Library

ISBN 978-1-78427-471-9 Pbk
ISBN 978-1-78427-472-6 ePub
ISBN 978-1-78427-473-3 PDF

Front cover main barn owl image © Ernie Janes/naturepl.com.
All other cover photographs © the author.

Printed in the Czech Republic by Finidr

To my loving and supporting wife Liz and wildlife-inquisitive sons Freddie and Benjamin.

CONTENTS

ABOUT THE AUTHOR

Ed is a passionate zoologist and has been following his love for wildlife for over 40 years. Alongside taking people out to see wildlife near in his home in the Forest of Dean, Gloucestershire, Ed has been a tour leader on adventures with guests helping them to see and learn more about wildlife all around the world. Ed has a passion for being outdoors and showing others all wildlife and their behaviour, helping to explain what's going on and experiencing it from a landscape perspective.

Ed loves studying animal behaviour and has been collecting bird pellets, skulls and feathers since he was seven years old. Since 1998 he has been researching what urban-dwelling peregrines eat. For four years Ed worked with over 200 students at the University of Bristol each spring, with each student dissecting barn owl pellets in the laboratory. The huge sample size each season allowed a remarkable insight into what barn owls close to Bristol and on Salisbury Plain were eating.

Away from his research work Ed is a naturalist, broadcaster and learning consultant engaging a wide range of people with nature and science. Activities range from tour guiding around the world, dawn chorus walks listening to birdsong, wildlife surveys, taking schools fossil hunting and developing learning resources and evaluating learning projects. He has done extensive work with the BBC as a contributor, consultant and reporter over the past 20 years, appearing on shows including the BBC's One Show, Springwatch and Autumnwatch and Radio 4 Natural History Radio.

Ed is the author of other Pelagic Publishing books: *Urban Peregrines* and *Raptor Prey Remains*.

1 INTRODUCTION

Dissecting bird pellets and seeing what is inside is the ultimate fun way of discovering what a particular group of birds has been eating. This is often an activity offered at family events or in schools by wildlife organisations – it has all the fascinating and disgusting elements of play that children love. Extracting information from pellets also has sound scientific value: while it does not capture everything a bird has been eating, this method still goes a long way in revealing the diets of birds and how this may change over time in different habitats or parts of the world. Pellet analyses may also help indicate the presence of species not otherwise detected in a habitat, or provide early indications of the decline or increase in populations of species, such as marine fish. They can also be used to look at the role birds have in the dispersal of seeds and fruits and to study what bird predators have eaten thousands or even millions of years ago.

Most birds in Britain and Ireland probably produce pellets, especially if they are eating anything that contains indigestible material such as hair, bone or the hard, external body parts from invertebrates. While birds of prey, including owls, eagles, hawks and falcons, are well known for producing pellets, even garden birds, such as robins and blackbirds will produce them. Other more surprising species include reed warbler, sedge warbler, spotted flycatcher, nightjar, woodpigeon and capercaillie (Tucker 1944).

In this guide I showcase pellets from a wide range of different bird species that are likely to be encountered and reveal how to identify common items found in them,

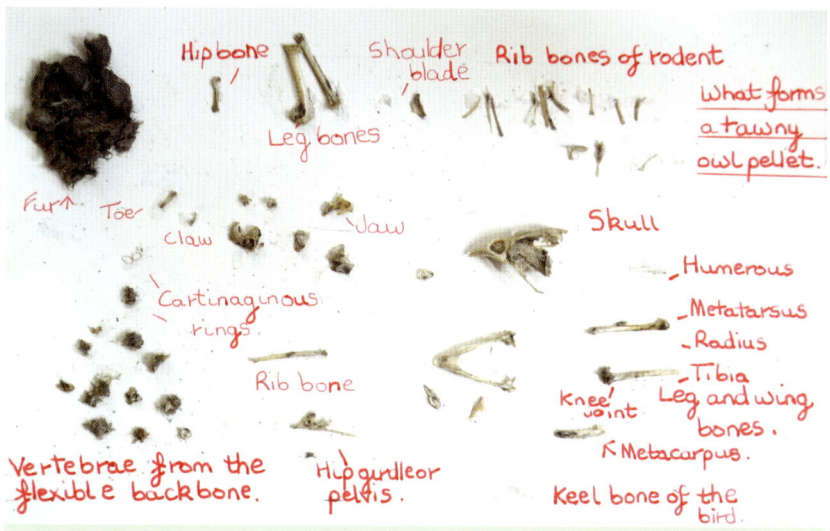

The skull and bones of a house sparrow from a tawny owl pellet that I dissected and laid out when I was a boy (Leatherhead, Surrey, 1990s).

such as small mammal skulls and bones. While the book focuses on pellets from bird species in Britain and Ireland, it also features some images that have come from the same species found in other parts of the world. Any images without a person's name in the caption were found or taken by myself. The scientific names of most species are summarised in a list at the back of the book.

As a boy I loved finding and collecting anything that came from birds – feathers, skulls and pellets. They gave me a connection and insights into different species that I was unable to get from a book or photos. Some of my first pellets were those from garden birds: a robin that was nesting in the garden and a sparrowhawk that had a favourite plucking perch in the woods near where I lived. Those original pellets are still safely laid out in small boxes on a layer of sawdust. In my late teenage years I started studying the diet of urban peregrines in Bristol and other cities and would regularly find their powdery, crumbly pellets. In the mid-2010s I had the pleasure of working in the teaching laboratory in the School of Biological Sciences at the University of Bristol. One of the practical sessions each spring was 'Mammals as prey'. Over 200 barn owl pellets would be dissected each year, revealing a brilliant dataset of what barn owls had been eating at Avonmouth Sewage Works, Bristol, Chew Valley Lake or Salisbury Plain. It was these practical sessions that particularly inspired me to look closer at bird pellets and discover what lies within them. I hope this book inspires you too to do the same.

2 WHAT IS A PELLET?

A bird pellet is the undigested remains of food that has been collected in the stomach, rather than passed through the alimentary canal and been excreted as poo. Instead, it is packed up into a pellet and regurgitated back up out of the mouth. Contrary to common belief, it is not just birds of prey such as owls, hawks, falcons and eagles that produce pellets. Most birds that eat substantially hard and indigestible parts of animals or plants in fact regurgitate pellets, including woodland and garden bird species such as blackbirds, song thrushes and robins, corvids like magpies and rooks, and seabirds such as gulls, terns, skuas and cormorants.

Depending on what a bird has been eating, its pellet may appear soft and hairy, or rough and textured where it is made up of the hard parts of seeds, insects or reptiles. Hairy or feathery examples may feel solid and hard to the touch thanks to an outer coating of mucus that seals the pellets, creating a more robust outer surface that deteriorates over time. In the case of seabirds such as cormorants, this mucus is present in abundance and solidifies like glue, turning pellets rock-hard.

A particularly varied and colourful pellet, probably from a magpie, jackdaw or crow, containing a mix of the hard outer skeletons of invertebrates including beetles, woodlouse and the wing of a cranefly. (Dr Phil Gates)

The quantity and quality of identifiable material found in pellets depends on how birds feed – whether they pick off food from prey items or eat them whole – and their ability to partially or fully digest prey. For instance, peregrines largely feed around the bones of a bird and often tear off wings, heads and legs without eating them. Therefore, their pellets comprise undigested feathers packed together with the occasional bird foot, bone fragment or bird identification ring. A barn owl on the other hand will eat a small mammal whole and regurgitate most of its bones and hair, although other owl species will digest bones more easily. Sometimes traces of flesh will still be apparent on skulls, feet and tails where they have not been fully digested. The hard parts of invertebrates, such as beetle wing cases (elytra), remain unscathed and just as shiny and bright as when they were swallowed.

Distinguishing a pellet from poo

Pellets are often confused with faeces or poo from other animals, in particular mammals that may consume similar prey. Generally, though, pellets contain parts of invertebrates, hair, feathers, bones, teeth and other hard body parts that are more or less intact – unlike in the case of mammal excrement. They haven't been crushed or mushed up by the hard, crunching teeth of mammal predators or scavengers. These pellets can be pulled apart with hands or tweezers and do not contain the soft, squidgy parts of poo that we see from mammals. They also tend to have a cleaner or musty smell.

Some mammals do produce pellet-like poo, depending on what they have been eating. When foxes have been eating birds or mammals, the feathers and hair are included in their faeces. These are long, thin and often twisted, tapering at their ends. They may smell musty and appear more like a dog poo rather than a bird pellet. The remains of any bones or parts of bodies such as feet are more likely to be chewed and fragmented, or indistinguishable. Squirrel poo may resemble small garden bird pellets. They look fibrous with partly digested plant material – a little like the texture of a rabbit or hare poo, although shaped like a small sausage. However, they lack the whole seeds or pips of fruit and invertebrate remains such as spiders or woodlice that are found in the pellets of songbirds.

The poo of birds is generally watery, white splashes with a darker solid mass. In pigeons these are somewhat more wholesome, soft and brown-green with white deposits of uric acid. Those of waterbirds, such as geese and swans, are soft, green and fibrous, while those of grouse are fibrous and dry. Birds of prey produce copious white splashes that whitewash favourite cliff ledges and vegetation below nests. The green woodpecker produces long, cylindrical poos packed full of the exoskeletons of ants encased on the outside with a thin layer of white uric acid.

A badger poo containing blackberries in the Forest of Dean.

Fox scat (poo) deposited by a rhyne (water channel) on the Gwent Levels.

Poo from a hedgehog – which is typically black (and compared next to a worm cast).

More hedgehog poo, showing its fibrous texture, often revealing crunched-up invertebrate remains such as tiny snail shell fragments.

Otter spraint (poo) smeared across a concrete step by a rhyne (water channel), showing tiny fragments of fishbones.

Mink scat (poo) left on a sea wall and revealing a tiny fish vertebra.

Pellet-like woodpigeon poo, with obvious ivy seeds and the typical white tip of uric acid.

A woodpigeon poo.

Another woodpigeon poo.

Typical poo of a bird such as a magpie – which this is most likely from – showing the darker liquidy solids and the white uric acid surrounding it (the fly appears to have subsequently drowned in the liquid).

The cylindrical poo of a green woodpecker, packed full of the exoskeletons of ants and showing the distinctive white end produced by uric acid.

Pellet sizes

Pellets come in all sorts of shapes and sizes. Each species of bird will produce pellets of varying size, and this can make identification more challenging as they do not always relate directly to the size of the bird they came from. For example, ravens produce both large pellets (80mm long) and small pellets (50mm long or less). Pellets from the same species vary in size depending on what they have been eating and on their age: young birds will produce small pellets relative to those produced by older individuals. Pellets that have been around for a while may also appear larger than when they were first regurgitated. For instance, barn owl pellets begin to expand and appear larger as they dry out and the outer mucus coating slowly breaks down. This process may be even faster if detritivores – invertebrates that love eating feathers and hair – get to work (see 'What else might you find in a pellet?').

How does a pellet form?

A pellet forms in the stomach of a bird. How much bone is regurgitated is determined by how much has been swallowed and by the digestive power of a given species' stomach juices. A bird's stomach is made up of two main parts: the proventriculus and the ventriculus or muscular stomach, which is known as the gizzard. The undigestible remains from a meal – seeds, hair, feathers, bones, teeth, horny parts, such as claws and hard parts of invertebrates – are formed together in the gizzard and stored in the smaller, weaker proventriculus prior to regurgitation.

In medium- to large-sized owls, pellets are formed within a six- to eight-hour period (Grimm & Whitehouse 1963; Smith & Richmond 1972). Pellets are regurgitated in response to how much food the bird has been eating, when it last fed and by the stimulus of hunting new prey (Smith & Richmond 1972). Large birds such as owls and cormorants produce one pellet per day, sometimes two, while skuas and gulls produce one per meal (Marti 1973; Barrett *et al.* 2007). By contrast, green sandpipers produce pellets every 20 minutes when feeding (Holt & Warrington 1996)!

Variable colour

Pellets vary in colour depending on the diet of the bird that produced them. Those of crows and rooks are very light and yellow in appearance if they have been feeding on grain, while an abundance of insects may result in black or brown pellets. For little owls and jackdaws, these may be black, glossy and iridescent due to the shiny beetle wing cases. Birds of prey produce grey, dark brown or almost black pellets when feeding on small mammals. The remains of voles, mice and rabbits tend to form grey-brown pellets – although they still tend to look black or dark brown in barn owl pellets – while moles and shrews (if eaten as the sole species) lead to black pellets. When invertebrates such as cockchafers, earthworms and spiders have been eaten, the pellets are usually browner – a mix of body parts, earthworm chaetae (hairs) and soil, especially sand. Those birds feeding on other birds tend to produce lighter coloured, often grey pellets, as a result of swallowing the soft, fluffier down feathers that are less easy to pluck. Seabird pellets are often creamy-yellow, formed of fishbones and the hard parts of marine invertebrates.

Finding pellets

When walking outdoors, including just in a garden or yard, there are often signs of wildlife from nibbled leaves, poo, cracked nuts or seeds and pellets. Pellets can be encountered anywhere that birds are likely to perch, rest or be foraging. In gardens, small songbird pellets may be found in feeding areas or beneath their nests. Colonies of nesting birds, such as bee-eaters, will leave pellets on the ground beneath their nest holes. Birds of prey and corvids (members of the crow family) frequently regurgitate pellets onto fenceposts, anthills or other resting perches. Barn owls leave many of their pellets inside their nests, often a tree hollow or a purpose-built nest box. On buildings, urban-dwelling gulls leave pellets on roofs near their nests and where they rest and loaf around. Winter roosts of birds, such as pied wagtails, starlings, ravens, rooks and jackdaws, that often overnight on building structures or trees, will leave pellets below their perches. Wader and seabird pellets may be found at favourite resting or roosting locations, such as offshore lighthouses, piers, rocks, breakwaters – but only at times that are not disturbing to individuals or flocks at high tide. There are even purpose-built floating pellet-collecting devices for birds such as cormorants to study their diet without disturbing or killing them (Gagliardi *et al.* 2003; Barrett *et al.* 2007)! By their nature cormorants like to rest in trees overhanging water in often inaccessible places. Deliberate floating platforms placed beneath such roosts helps catch regurgitated pellets as they fall towards the water below.

While pellets are often found incidentally when out and about, others may need to be deliberately collected during arranged visits to known roost sites – during the period when the birds are away hunting or foraging – and at or below nesting locations. Depending on the species, some are best visited with or by established bird of prey or seabird groups who have the required experience and licences and may be willing to collect on your behalf. For some species, disturbances to the nest also require a Schedule 1 licence and the collecting of pellets ideally needs to be done alongside collecting nest data on eggs and young, or if the young are being ringed (where they have a unique metal identification ring fitted on their leg by volunteers on behalf of the British Trust for Ornithology (BTO)). The list of Schedule 1 species can be found at legislation.gov.uk/ukpga/1981/69/schedule/1 and on the Royal Society for the Protection of Birds (RSPB) and BTO websites.

Barn owl pellets are often regurgitated onto the ground below a perch or nest box, making them easy to find.

Pellets are often found on top of fenceposts and gates. This corvid pellet was on top of a gate.

Corvid pellet found on top of a gate post on the Gwent Levels.

This older owl pellet (tawny owl or barn owl) was among twigs and leaves in the open hollow of an ash tree.

Other pellets may be encountered alone on the ground – this one, from an owl or buzzard, was on a log beneath an oak tree.

Ageing pellets

Many pellets are very delicate. When first regurgitated they are wet and soft, often coated in mucus and easily disintegrate. Small songbird pellets are more prone to falling apart, especially when fresh. As pellets dry out some become more brittle and will fall apart, while others will adhere together more easily. Those of cormorants and shags may dry hard like glue and be easy to handle. Pellets full of animal fur or feathers will hold together well and expand as they dry out. Older pellets will often reveal their contents as the outer layers decay and fall away.

The *Barn Owl Conservation Handbook* and the Barn Owl Trust website provide photos of the different stages of the decay of barn owl pellets in a sheltered roost or nest site, from a fresh pellet to one that has been lying there for 30 months or longer (Barn Owl Trust 2012).

These two pellets are old and weathered, revealing their contents more clearly rather than being wrapped up in hair, feathers and mucus. The first pellet reveals a matrix of frog bones, while the contents in the second pellet (collected by Jason Fathers) are made up of fishbones.

The scattered bones of small mammals, all that remain of barn owl pellets that have largely rotted away.

Corvid pellets may also be found in various stages of decay, such as these crow or raven pellets on a stone wall. Those that drop down into crevices are hidden from the worst of the weather and may remain intact for many months. (Bob Cowley)

A pellet, full of dung beetles, that has disintegrated. (Bob Cowley)

A similar type of pellet just in the early stages of breaking apart. (Bob Cowley)

Remains from meals of prey: secondary consumption

When dissecting bird pellets you might also find the remains of food that their prey has eaten before being hunted and swallowed. For example, peregrine and goshawk pellets may contain wheat grain from the crops of pigeons they have been feeding on. Sparrowhawks may feed on songbirds that have seeds and berries inside their stomachs. Cormorants may feed on fish that themselves have been feeding on smaller fish, shellfish or worms. There may also be parasitic animals inside or on the outside of prey that is swallowed. Studying this secondary consumption of items in pellets can help explore how bird predators contribute to the dispersal of seeds or invertebrates, including parasites, and their eggs (van Leeuwen *et al.* 2017; Navarro-Ramos *et al.* 2022; Green *et al.* 2023). Lead and non-lead shot may also be found in pellets of birds feeding on carrion or live birds they have predated that have ingested shot, been killed by shot or were predated with shot already embedded in their bodies. These include starlings, pigeons, gamebirds and waterbirds (Kendall *et al.* 1996, Mateo *et al.* 1999; Mateo *et al.* 2001; Andreotti & Borghesi 2013).

Plastic particles

Plastic particles are frequently found in bird pellets, including those of gulls, skuas, terns, cormorants, dippers, kingfishers and corvids. Although public awareness of plastic being eaten by wildlife has increased in recent years, ornithologists have been finding plastic particles in pellets for decades. For instance, tiny pieces of polystyrene known as spherules – from industrial waste – were found in gull and tern pellets in New York in the early 1970s (Hays & Cormons 1974). More recently there has been concern at the amount of microplastics found in the pellets (and poo) of freshwater birds such as dippers and kingfishers, particularly as the effects they may have on the birds are still poorly understood (D'Souza *et al.* 2020; Winkler *et al.* 2020).

Other litter may also be found in bird pellets, particularly those of gulls, ranging from glass to kebab skewer sticks and condoms to plastic bags. Great skuas are thought to swallow plastic litter as secondary consumers when they eat fulmars and other seabirds (Hammer *et al.* 2016). Recent research on barn owls has also revealed microplastics in their pellets, thought to be transferred from small mammal prey they are eating (Nessi *et al.* 2022).

3 DISSECTING YOUR OWN PELLETS – WHERE TO BEGIN

With family and friends at home or at events

Owl pellets are generally the easiest pellets to get hold of, either from local bird of prey (raptor) groups or from organisations such as Raptor Aid and the Barn Owl Trust (who can supply a barn owl pellet dissection pack). If you receive them damp they are best stored in a freezer until they are being used, or be allowed to dry and then stored in a cardboard box or envelope to avoid them going mouldy.

Dissecting owl pellets can easily be done indoors and outdoors, at home, in a village hall or at a festival, or in an education setting. It is a brilliant way of engaging others with nature, particularly children and young people, and helping to make connections between owls, their food and the environment they live in. At events or organised activities, having a captive owl from a reputable bird of prey business enhances the experience and gives greater relevance to the pellet activity.

What you need

- A white tray, glass dish, pale-coloured washing up bowl or ice-cream tub – something that will contain the pellets' contents and water and have a contrasting, lighter background so you can spot things.
- Disposable gloves.
- Tweezers.
- Water.
- Absorbent paper to dry bones.
- Magnifying glass or microscope (if available, not essential).

Keeping healthy

Some bird diseases can be passed to people and cause illnesses. Although pellets pose a low risk, it is still best to minimise any hand-to-mouth contact or inhaling dust from them by:

- Risk assessing any activity or event with hygiene in mind.
- Soaking pellets to reduce any airborne dust rising up from the pellets and being inhaled.
- Keeping hands away from mouths until washed to avoid swallowing any germs.
- Washing hands with soap and water (and not just hand gel) to get rid of any bacteria and other germs.
- Providing disposable gloves for those who wish to use them.

1. Owl pellets are a well consolidated mass of hair and bones. They will usually be dry and hard when supplied and can be broken up dry. However, wetting them enables all the hair and bones to be teased apart more easily. Disposable gloves can be worn, although sometimes it is easier to handle pellets and bones with bare hands and wash them well afterwards.

2. A pellet can be put into a tray or dish and some water added – enough to soak the pellet. Leave for a few minutes. As the pellet absorbs the water it will become softer and can then be opened using fingers or tweezers.

3. Pick out skulls and other bones and place on paper or a separate tray. The small mammal hair will mix with the water and turn it into a dark, hairy non-edible soup! However, any bones and insect remains can be spotted or felt and placed onto dry paper or a tray.

4. Some pellets may contain no skulls, some just one and others may contain several (perhaps even four or five) alongside other bones. The lower jawbones usually split apart; there are usually the separate left and right jawbones for each skull somewhere in the same pellet, although the bones of one animal may be split between several pellets.

5. Once all the hard bony remains are laid out, identification and counting can begin! The best way to do this is to display all the same skull types together (upside down so their teeth are showing). Do the same with the lower-left and lower-right jawbones. Each different species can then be identified (using this book) and individuals counted.

6.   Depending on the nature of the setting or event, repeat the process with other pellets to build up a list of prey items. These can be used to make a bar chart or pie chart outlining what has been found.

You might also:

*   Photograph or draw your findings.
*   Find out more about the biology of the prey animals eaten.
*   Report any interesting findings online to your local biological records centre who collate recordings of nature across counties or regions in Britain. They are interested in both common prey items and more unusual species including water shrews, water voles and harvest mice. Your findings may reveal new or unknown locations for such animals. To find your local centre search on the Association for Local Environmental Records Centres website, alerc.org.uk/lerc-finder.html.

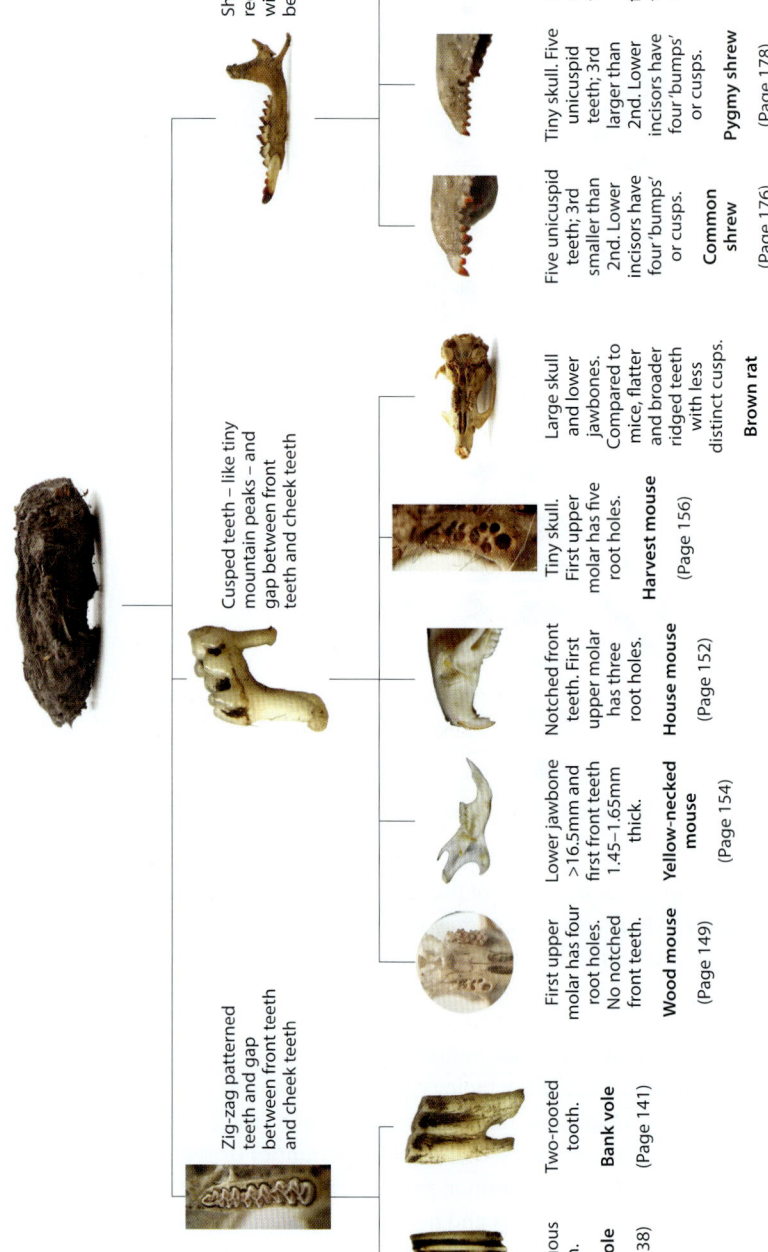

Owl pellet

Zig-zag patterned teeth and gap between front teeth and cheek teeth

Continuous tooth.
Field vole
(Page 138)

Two-rooted tooth.
Bank vole
(Page 141)

Cusped teeth – like tiny mountain peaks – and gap between front teeth and cheek teeth

First upper molar has four root holes. No notched front teeth.
Wood mouse
(Page 149)

Lower jawbone >16.5mm and first front teeth 1.45–1.65mm thick.
Yellow-necked mouse
(Page 154)

Notched front teeth. First upper molar has three root holes.
House mouse
(Page 152)

Tiny skull. First upper molar has five root holes.
Harvest mouse
(Page 156)

Large skull and lower jawbones. Compared to mice, flatter and broader ridged teeth with less distinct cusps.
Brown rat
(Page 158)

Sharp red-tipped teeth with no gaps between teeth

Five unicuspid teeth; 3rd smaller than 2nd. Lower incisors have four 'bumps' or cusps.
Common shrew
(Page 176)

Tiny skull. Five unicuspid teeth; 3rd larger than 2nd. Lower incisors have four 'bumps' or cusps.
Pygmy shrew
(Page 178)

Similar size or slightly larger than common shrew. Only four unicuspid teeth. Smooth lower incisors.
Water shrew
(Page 180)

- Look at the skulls under an illuminated microscope to see more of their detail.
- Produce a report-style document outlining what you or your group have done, especially if in a secondary or university learning setting.

Basic guide to identification

The diagram opposite (page 24) offers a quick key to the teeth and skulls of the most common small mammals found in owl pellets on mainland Britain (for other species see specific profiles in this book).

Top tips to quickly identify small mammal skulls include:

- Voles have zig-zag patterned teeth, while mice have cusped teeth with tiny ridges.
- Generally, the teeth of bank voles have two roots while those of field voles have one root. To check, use tweezers to pluck teeth out of vole skulls or lower jawbones.
- For mice, the number of root holes in the teeth closest to the brain case or cranium can help with identification. Teeth can be plucked out using tweezers.
- Common, pygmy and water shrews have red- or orange-tipped teeth.
- Rat skulls and lower jawbones, even from young animals, will appear much larger than the vole or mice skulls or jawbones.

Schools

Schools are an ideal setting for pellet dissections. You can usually source enough pellets for each learner to have one, or for working in pairs. If you do not know of any local bird of prey groups or owl ringers, see if your local wildlife trust has any contacts or try organisations such as the Barn Owl Trust and Raptor Aid which supply pellets. Pellets that arrives from these sources are likely to be dry and can be stored in a paper envelope or box. If they are damp either use them straight away, allow them to dry outdoors on newspaper or on a tray, or put them in a freezer (and defrost them the day before use).

Children (and adults) get really engaged and absorbed in this activity so it is worth allowing at least one or two hours.

There are a wide range of ways in which dissecting pellets complement the curriculum, including:

- Creative writing about barn owls and their prey – what happened before the small mammals were eaten?
- Exploring what an owl needs as a healthy diet, perhaps presented as a menu.
- Discussion of food chains and food webs.
- Considering the adaptations of predators and their prey.
- Comparing small mammal bones with those of humans.

- Habitats – where do barn owls and small mammals live?
- Using the data on animals found in the pellets to create a dataset and present as pie charts, bar charts and percentages. Other data could include the length and weight of pellets before dissection and how many animals are found per pellet.

WildlifeKate (Kate MacRae) has been involved in the primary sector for over 30 years, enthusing and exciting children and educators about the natural world and about the benefits of being in the great outdoors. She now supports schools and many other organisations as a consultant, alongside lots of exciting wildlife-camera projects across Britain.

Owl pellet dissections remain her favourite classroom activity and here she explains how teachers can make the most of this amazing learning experience.

"I usually link barn owl pellet dissection in with food chains, habitat or adaptations work. I have done this activity from key stage 1 all the way through to top key stage 2. Although there is a little trepidation at the start, this soon dissipates as the bones and skulls start to emerge!

I tend to start the lesson with clips of barn owls and information about them. I often use clips from YouTube of barn owls, so the pupils understand what the barn owl is, how it lives and how it hunts. If possible, I show them a stuffed barn owl or even better, get a local bird of prey display group to bring a live barn owl into the classroom.

Regarding the health & safety aspect: I soak the pellets in a little warm water with disinfectant a few minutes before giving them to the kids. I usually have tweezers or I break wooden kebab skewers in half which they can use to separate the pellets. However, using hands to break up the pellets and tease out bones is far easier and so that is what I model! I obviously talk about

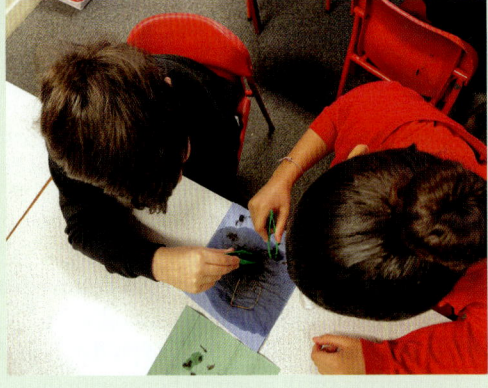

hygiene and that we all wash out hands thoroughly afterwards. Many kids start with the tools but then go to hands as it is much more fun!

I give them two paper towels. One is for the pellet and the other is to lay out the bits they find. I use a variety of resources, mainly bits I have printed and got from the internet so they have an idea as to what they might be finding. We generally try to ID vole/mice and shrews and some of the other easily identifiable bones. The skulls are the things they get really excited about though! I show them the skull and the two sections of the lower jaw and show how

they fit together and explain why they are separate in the pellet. We often get maggots in the pellets as I get them from all sorts of wild sources. I tell the kids these are special caterpillars of flies and moths that help break down the pellets and that they can put them to one side (in a pot if still alive) and we put them out for the birds. The staff freak out if you say they are 'maggots'!! [for more information on these

larvae see 'What else might you find in a pellet?' later in this book]. I also get as much magnifying equipment for the kids to try as possible. Digital microscopes and visual-isers that can project onto the whiteboard are useful too.

I get the kids to put any bits they want to mount onto the clean paper towel. I encourage them to try and get them as clean as possible by pulling off all the fluff and dirt. Each child has a piece of black card. They place a blob of PVA glue on the card and sink the skull/bone into it. It dries clear. We look at the range of species found and think about what that tells us

about a habitat. They usually want to take them home with them, although I have done displays in the past and linked it to literacy and science work, researching the mammals and finding out more about them and their habitats."

Photos by Kate MacRae, wildlifekate.co.uk

Student practicals

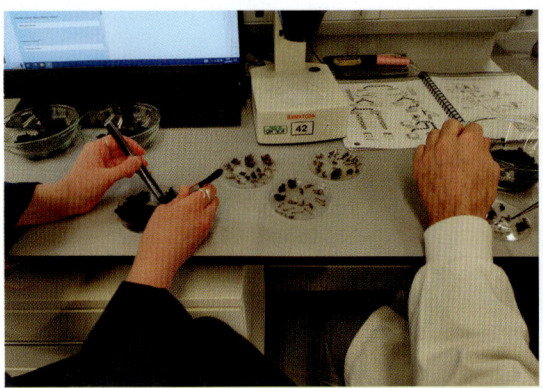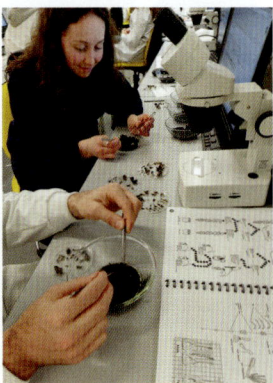

An owl pellet dissection session as part of an undergraduate practical at the University of Bristol. (Sue Howell)

Owl pellets are an ideal subject for post-16 or undergraduate student dissections, particularly at colleges and universities, when looking at predator–prey relationships. They are also an opportunity for data analysis, producing charts or graphs and writing reports. In such cases, pellets may be needed in their hundreds and can be sourced from local bird of prey groups or ringers involved with owls, or ecology consultants that may be visiting nests (under licence or out of season).

Student practicals generate lots of data and big samples, and the data can be fed back to those who supplied the pellets. Interesting findings should be reported to the county biological records centre – for example, if remains of scarcer species such as harvest mice, water voles or water shrews are found. Flying a captive owl to complement the practical also enhances the learning experience.

As well as recording species and their total frequency, the number of individual animals found per pellet – determined by the number of skulls – alongside the pellet's length, width and weight can also be measured. The pelvic bones of small mammals can help determine the ratio of male and female animals (see 'Sexing small mammal bones'), while the length of the humerus bones of birds can be used to estimate the bird's weight, even if its identity remains unknown (see 'Bird bones (excluding skulls)').

Research: using pellets in diet or prey studies

Pellets are just one means by which the diet of birds can be studied. Web cameras and direct observations of food deliveries can be a useful tool for seeing what birds, in particular birds of prey, are eating during the breeding season. However, while they show what is being brought directly to the chicks, they do not reveal everything the adults may be eating away from the nest or at other times of the year.

The study of bird pellets is most likely to provide an indication of what particular birds are eating (diet composition), rather than an accurate indication of how many individuals of different species are consumed (diet quantification). Pellet analyses more likely reflect the number of days on which such prey was eaten, or provide an index of abundance rather than actual number of prey type (Village 1990). Pellets also offer data that is biased towards larger and more obvious prey; smaller prey items from invertebrate parts or small mammal bones, or those less easily identified, may be missed or are less likely to be logged.

Despite these biases, dissecting bird pellets and identifying what is inside is an important tool for discovering what birds are eating, either in isolation or combined with other methods, such as collecting prey remains and direct observations. They are also a great way of collecting data while causing minimal disturbance to the bird species involved. However, the amount of prey you find in pellets will vary between species. Different birds digest hard parts, such as bones, in different ways – so, even if bones are swallowed they may not come back up in a pellet.

What goes in isn't necessarily what comes back up!

For some bird species, such as barn owls, small mammal prey is swallowed whole with little sign of them – such as fur or body parts – left behind. Therefore, their pellets may be the only means by which we can find out what they have been eating. Alongside bird or mammal skulls, the numbers of lower jaws, mammal pelvic bones and bird breastbones and pelvises can help provide a more accurate count of what has been swallowed. At times there may also be wings, legs and heads that have been removed by a predator and remain uneaten nearby (Glue 1977). For example, barn owls will often leave uneaten wings, legs and heads of water rails and starlings in their nest or roosting sites and thus offer clues to what else they have been eating. Grey herons, on the other hand, digest most bones from small mammals; their pellets largely comprise hair and occasional teeth, meaning identifying the species and numbers of prey eaten is more difficult. For other species, pellets form just a small part of the story and finding out more about their diet often depends on the way in which that species eats its prey. Peregrines, for instance, tend to pick around a bird carcass rather than swallowing many bones and hence their pellets mainly comprise small, softer body feathers. Because of this, looking for prey remains, such as wings, legs and heads, is more important in studying the diet of peregrines – and pellets are complementary.

In summary, while some species may digest much of their prey (including bones), others may digest just the soft tissues and regurgitate any hard parts as pellets. For some birds, small hard parts may pass through the whole digestive tract and emerge in their poo, or stay in their stomachs for many weeks or months before reappearing in pellets or poo. Therefore, other methods can help build a more accurate and less biased picture of what a bird has been eating. These might include examining the poo in detail, particularly for birds such as waders and dippers, and collecting undigested regurgitated food for species such as gulls. Despite what can be missed by solely using pellets to determine the diet of birds, comparative studies over time and across different locations and periods can reveal interesting and important insights.

DNA metabarcoding and isotope analyses

Both the use of metabarcoding from environmental DNA in pellets and stable isotope analyses are now available to researchers in order to help establish what a species has been eating.

Metabarcoding enables DNA extracted from the pellets to be compared with known DNA reference sequences of different species. This allows both the producer of the pellets to be identified (including its sex) alongside what it has consumed (van der Reis & Jeffs 2020; Hacker *et al.* 2021; Shimizu *et al.* 2022). This approach also facilitates genetic studies on the prey species, such as small mammals (Guimaraes 2016).

Isotope analyses, on the other hand, look for the ratio of stable isotopes of elements such as carbon and nitrogen, which differ between species depending on what they eat and where they live. This is known as their isotope signature or pattern. Specific ratios of stable isotopes present in the feathers or muscle of the bird producing the pellets indicate particular prey species that may have been missed through pellet dissections because it was digested or too small to see or identify. When the prey is digested, those specific isotope ratios get incorporated into the bird's feather structure or muscle and will differ from the bird's own isotope signature (Li *et al.* 2004; Weiser & Powell 2011; Resano-Mayor 2014; Catry *et al.* 2016). These analyses can be compared with conventional methods such as dissecting pellets, although isotope analyses may reveal the diet over a longer period of time. For example, a feather may stay on a bird for a year or more and its isotopes will reflect what the bird ate while the feather was being grown. A pellet, on the other hand, will reveal what a bird has eaten just hours or days before it was regurgitated.

Pellets can also be useful in taphonomy – the study of how animals decay and fossilise or are preserved over time – and may provide useful clues to palaeontologists and archaeologists on what animals once lived in an area by looking at accumulations of small animal bones. Studies to date have attempted to analyse how owl pellets decay, the characteristics of their contents (such as broken bones, amount of bone and tooth loss by digestion), what happens after they are regurgitated (such as weathering and becoming buried) and how they may be distinguished from different owl species and other predators that would have been living in similar places and feeding on similar prey in the past (Dodson & Wexlar 1979; Kusmer 1990; Laudet *et al.* 2002; Terry 2004; Hockett 2018; Lopez 2020).

4 OWL PELLETS

Owl pellets are easily identifiable and commonly encountered, with those from barn owls being the most often dissected by children, young people and adults at family-friendly wildlife events. This is partly because their pellets can be collected – under licence – by surveyors in bulk at many nest locations where they accumulate, and partly because a lot of effort is invested in monitoring nest boxes used by barn owls. Their pellets are large and when soaked in water generally reveal anywhere between one and five (perhaps more) small mammal skulls and other bones.

Different species of owls produce very similar-looking pellets and if they were to all be mixed together there would be few differences enabling us to tell them apart, although barn owl pellets are particularly large and dark. This is because many share similar prey, albeit in different proportions depending on the owl species, their habitat and differing prey populations depending on where they live. Barn, tawny, long-eared and short-eared owls all feed on small mammals, small birds (to a greater or lesser degree) and may consume other animals too, such as frogs and insects. The little owl produces smaller pellets mostly dominated by invertebrates, although small mammals and birds do feature. Therefore, the identification of owl pellets is usually determined by the location and situation in which they were found and are often accompanied by the tell-tale feathers from the owls themselves. Sometimes, though, the species of owl they came from may remain a mystery. There can be overlap in the habitats used by owl species and there may not be other signs, such as feathers, to confirm identity. For example, tawny owls and barn owls may use similar trees along the edge of woodland and hedgerows. Generally, nest monitoring and the use of artificial nests such as boxes are good ways of knowing for sure who the pellets are from, as the subsequent nest visits to record eggs and chicks, as well as ringing activities, will confirm the species of owl using the box.

Note: All images of pellets and pellet contents are shown life size unless a scale bar or a note in the photo caption indicates otherwise. Outdoor, 'in-the-field' images are not to scale.

A typical large, black-brown pellet from a barn owl found at a nest location. (Collected by Colin Morris)

Barn owl

The white, watery poo of barn owls is sometimes a giveaway to their presence. At this location the white poo led to the discovery of pellets at the base of the oak tree.

An accumulation of decaying pellets (revealing their contents) over many seasons in a nest box.

Some pellets may be regurgitated on the outside ledge of a nest box.

Barn owl pellet. (Collected by Gordon Kirk)

A selection of barn owl pellets. (Collected by Colin Morris)

Barn owl pellet containing the head of a starling. (Collected by John Boorman)

Diet: Mostly small mammals, although may include birds such as starlings, yellow-hammers and water rails and occasionally frogs. More likely to eat shrews compared with tawny and long-eared owls.

Identification: Black-brown thick bulky pellets varying in terms of size and how compacted they are. Some are long and cylindrical while others are short and more rounded. When fresh, dried mucus creates a darker, shinier surface. Older pellets are greyer, less shiny and with more obvious bones at the surface. Creamy-white poo may cover some pellets.

Where found: In nest boxes, natural nest cavities such as tree hollows and on the ground beneath nests and resting (roosting) locations. Visiting active barn owl nests requires a special licence (e.g. Schedule 1) or permit. The pellets often fall down through tree hollows to the base of a tree. Where barn owls are nesting in barns and other outdoor buildings, pellets will accumulate on the ground beneath perches and nest boxes.

Size: Large pellets compared to other similar-sized owls, ranging from 29–83mm long by 15–40mm thick (Mikkola 1983; Barn Owl Trust 2012; Brown *et al.* 2021).

Notable features: Accumulation of large, dark bony pellets made from small mammal hair and bones of small mammals. Where barn owls feed by estuaries and mudflats, pellets may contain the long legs of wading birds (Stuart & Stuart 2013).

Tawny owl

A selection of tawny owl pellets, revealing the remains of cockchafers, frogs and small mammals.

Tawny owl pellet containing the remains of small mammal prey. (Collected by Anna Field)

Tawny owl pellet found after recent rainfall and which had started to decay, revealing the small mammal bones inside.

Diet: Small mammals, insects (especially beetles), amphibians (especially frogs) and small or medium-sized birds, ranging from great tits and chaffinches to blackbirds and thrushes.

Identification: Similar in size to barn owl pellets, although more narrow, greyer and less compact, with more variable prey contents and often tapered at one end. Similar or identical to those of long-eared owl, with which this species shares a similar habitat. The tawny owl is absent from the island of Ireland.

Where found: Less easy to find compared to barn owl pellets as many are regurgitated away from roosts and nests, often at favourite pellet-dropping perches which switch between seasons. However, known roosts, nest hollows and nest boxes are still worth checking. When exposed to the weather, pellets may disintegrate over weeks or a few months, lasting longer where sheltered from rain (although may be slowly eaten by moth larvae) (Mikkola 1983). Visiting active tawny owl nests may require a special licence (e.g. Schedule 1) or permit, e.g. on the Isle of Man.

Size: 25–84mm long by 10–28mm thick (Mikkola 1983; Brown *et al.* 2021).

Notable features: Depending on the season, tawny owl pellets may contain frog bones and invertebrate remains such as those of dung beetles and cockchafers. While the bony remains of small mammals do appear in tawny owl pellets, some also get digested (Mikkola 1983).

Long-eared owl

A selection of long-eared owl pellets from a summer roost. (Collected by Alan McCarthy)

A selection of long-eared owl pellets from a winter roost. (Collected by Ian Buxton)

Diet: Small mammals and birds.

Identification: As with tawny owl and short-eared owl, the pellets are greyish, cylindrical and tapered. Long-eared owls share similar woodland habitat to tawny owls, while the edges of forests may be shared with short-eared owls and barn owls. On the island of Ireland – where the tawny owl is absent – the long-eared owl may be found in both rural and urban locations.

Where found: In and beneath nests and below winter roosts. Visiting active long-eared owl nests may require a special licence (e.g. Schedule 1) or permit.

Size: 19–77mm long by 11–25mm thick (Mikkola 1983).

Notable features: Look similar or identical to other owl species. Known roosts and nest locations are best way of being sure of identification, alongside the presence of moulted feathers.

Short-eared owl

A short-eared owl regurgitating a pellet, Taiwan. (Jeff Lin – Flickr profile: 賞景者 Jeff Lin)

Short-eared owl pellet. (Collected by Mike Price)

Short-eared owl pellets found in their natural state, Malta. (Nicholas Galea, Birdlife Malta)

A range of short-eared owl pellets collected together, with a tell-tale moulted feather, Malta. (Nicholas Galea, Birdlife Malta)

Diet: Small mammals (including rabbits and occasionally weasels and stoats) and birds such as dunlin, starling and meadow pipits (Glue 1977).

Identification: Similar diet to barn owls, although appear shorter and greyer in colour. Can be confused with hen harrier and long-eared owl. Longer compared to hen harriers and when handled feel bony as the bones are closer to the surface (Holt *et al.* 1987). Of all the owl species, if pellets are found in open areas away from trees, they are most likely to be from a short-eared owl.

Where found: In and around nests and favoured resting/roosting perches, often on the ground. Visiting active short-eared owl nests may require a special licence (e.g. Schedule 1) or permit.

Size: 22–119mm long by 11–30mm thick (Mikkola 1983; Holt *et al.* 1987; Brown *et al.* 2021). Often small, round pellets roughly 10mm in diameter (Glue 1977).

Notable features: The poo found near or with pellets can be helpful to tell if they belong to a short-eared owl. Short-eared owls have buff or creamy poo with a black solid bead, while hen harriers have white to white/green poo which may have some blackish solid parts (Holt *et al.* 1987).

Little owl

Little owl next to a recently regurgitated pellet. (Graham Parry – Flickr profile: GrahamParryWildlife)

A little owl pellet packed full of beetle remains. (Collected by Gordon Kirk)

A little owl pellet revealing the remains of plant material, grit and some beetle remains. (Collected by Gordon Kirk)

A little owl pellet revealing the hair and bones of small mammals. (Collected by Gordon Kirk)

Little owl pellets revealing the remains of plant material and beetle remains. (Collected by Richard Clarke)

A little owl pellet revealing the remains of plant material and some beetle remains. (Collected by Anna Field)

A little owl pellet revealing the hair and bones of small mammals. (Collected by Anna Field)

Diet: Insects (especially dung beetles), earthworms, small mammals, birds and small reptiles.

Identification: During the summer little owl pellets are full of the remains of insects, in particular beetles, soil and plant material (swallowed alongside earthworms and other invertebrates). During the winter pellets are more likely to contain the remains of small mammals and birds, alongside earthworms.

Where found: In nest boxes or on the ground beneath nests. If nesting on the ground, close to entrances. May be found on favourite perches such as gravestones, logs or ledges of old buildings. Visiting active little owl nests may require a special licence (e.g. Schedule 1) or permit, e.g. on the Isle of Man.

Size: Similar in size to kestrel, 15–40mm long and 10–15mm thick (Brown *et al.* 2021).

Notable features: Little owl pellets glint and shine blue-green from the iridescent shiny black body parts of dung beetles. Corvids such as crows, which also feed on dung beetles, produce larger, longer pellets. Winter pellets are less distinct and contain more soil due to the birds feeding on earthworms.

Eagle owl

Eagle owl pellets found at their resting and feeding locations, Sauerland, North Rhine-Westphalia, Germany. (Gerd Kistner)

Three eagle owl pellets found together at a roost, Sauerland, North Rhine-Westphalia, Germany. (Gerd Kistner)

Older, decaying eagle owl pellets, Sauerland, North Rhine-Westphalia, Germany. (Gerd Kistner)

An eagle owl pellet, 118mm long, revealing some large bones, Sundern, Sauerland, North Rhine-Westphalia, Germany. (Martin Lindner)

Diet: Hugely varied. Predominantly mammals, from small voles to pine martens, hedgehogs, foxes, badgers and young deer. A range of birds, especially seabirds and ducks, alongside other owls, buzzards and peregrines. Reptiles, amphibians, fish and invertebrates (from beetles to crabs) also eaten occasionally (Mikkola 1986).

Identification: Large, thick bony pellets described as, 'compressed, irregularly cylindrical or conical' (Mikkola 1983).

Where found: Daytime roosts, plucking posts or nest sites.

Size: 30–178mm long by 19–42mm thick (Mikkola 1983; Brown *et al.* 2021).

Notable features: When feeding on frogs or fish, the subsequent pellets – often smaller – may contain vegetation such as grass, moss or heather which eagle owls are thought to deliberately swallow to help bind the bones in the absence of hair or feathers (Mikkola 1983). At the nest the abundance of pellets form a thick carpet prior to chicks leaving. Hedgehog spines may also feature in pellets.

Snowy owl

A snowy owl pellet from different angles revealing rabbit hair and bones, North Uist. (Collected by Jacqueline Tonkin, and kept at Bristol Museum & Art Gallery)

Older, decaying snowy owl pellets with protruding bones found on the ground, Vinalhaven, Maine, USA. (Kirk Gentalen)

Diet: Small to medium-size mammals from voles to rabbits; small to medium-sized birds including finches, buntings, young waders, grouse and ptarmigan. May occasionally take insects, frogs and fish (Mikkola 1986).

Identification: Variable in size and shape, packed full of large bone fragments and hair or feathers.

Where found: On the ground at daytime roosts and at nests.

Size: 52–120mm long by 25–43mm thick (Mikkola 1983; Brown *et al.* 2021).

Notable features: Pellets often contain snowy owl feathers that have been moulted and subsequently eaten. Eating their own feathers probably helps with binding undigestible food and forming a pellet (Tulloch 1968; Hardey *et al.* 2013).

5 FALCON, HAWK, KITE, HARRIER AND EAGLE PELLETS

Alongside owls, other birds of prey also eat a wide range of animals whose bones, hair, feathers or exoskeletons are regurgitated in pellets. However, compared to owls, their pellets may contain less bony material due to their nature of teasing flesh off bones rather than swallowing prey whole. Ospreys are not featured in this book as they mainly swallow soft flesh and bones that easily get digested and so pellets are rarely encountered (Häkkinen 1978). Therefore, to date any published research discounts using pellets in osprey diet studies (Francour & Thibault 1996). Wild honey-buzzards also do not produce pellets (Bijlsma 1999).

Note: All images of pellets and pellet contents are shown life size unless a scale bar or a note in the photo caption indicates otherwise. Outdoor, 'in-the-field' images are not to scale.

Peregrine

An adult peregrine regurgitating a pellet onto the stonework of a historical mill. (Ian Bradley – Flickr profile: Ian Bradley)

An adult peregrine regurgitating a pellet from one of the towers of Lincoln Cathedral. (Peter Taylor)

A typical peregrine pellet found on the ground beneath a church.

A peregrine pellet full of starling body feathers.

A peregrine pellet packed full of body feathers from a little grebe.

Peregrine pellets containing the feathers of Manx shearwaters.

A peregrine pellet revealing the crop of a bird, most likely from a pigeon.

Peregrine pellets full of green feathers from ring-necked parakeets.

A peregrine pellet containing feral pigeon feathers and plant material.

Another peregrine pellet containing feral pigeon feathers. (Collected by Dave Morrison)

A further peregrine pellet containing feral pigeon feathers.

A peregrine pellet containing male teal feathers. (Collected by Gareth Jones)

Diet: Birds, ranging in size from goldcrests to small to medium-sized ducks and small geese.

Identification: Round to cylindrical loose pellets mostly comprising small body feathers. Variable in colour depending on prey consumed. Some may be very white or pale grey, while others may be dark grey, especially when wet. Sometimes contain parts of legs and toes, bills and bird identification rings.

Where found: Beneath favoured resting places on buildings and trees; also found in and around the nest. Visiting active peregrine nests requires a special licence (e.g. Schedule 1) or permit.

Size: Quite variable, 16–87mm long by 10–30mm thick (Hardey *et al.* 2013)

Notable features: Generally very pale, greyish pellets as contain mostly pale pigeon feathers. However, may be darker if they have eaten a starling, or green if they have eaten a parakeet. Break apart very easily when dry or during very heavy rain; many pellets easily turn to powder when rubbed between fingers.

Kestrel

Kestrel pellets that have been regurgitated and dropped below a perch on a bird hide.

A kestrel pellet containing the scales of a reptile.

A close-up of the reptile scales.

A selection of kestrel pellets containing small mammal remains. (Collected by Jane and Steve Gilliard)

Diet: Mostly small mammals. Some small birds, reptiles and invertebrates depending on habitat and the season. May steal food from other birds of prey such as barn owls.

Identification: Small, grey and slightly tapered on some pellets. Contains less bony material compared to owls.

Where found: In, around and below the nest and roosting perches, which may include church porches, open barns, balconies or ledges of buildings and trees. Visiting active kestrel nests may require a special licence (e.g. Schedule 1) or permit.

Size: 20–40mm long and 10–25mm thick (Village 1990; Brown *et al.* 2021).

Notable features: Mostly contains hair, feather, hard parts from invertebrates or reptile scales. May contain some bones, although small bones may be completely digested. Kestrels often pluck the feathers and hair and tease off the flesh from the bone rather than eating a whole intact animal (Village 1990).

Merlin

A range of merlin pellets containing meadow pipit feathers. (Collected by Kim Leyland)

A merlin pellet containing pipit or lark feathers. (Collected by Jack Ashton-Booth)

The same merlin pellet rotated to reveal the leg of the prey. (Collected by Jack Ashton-Booth)

A collection of pellets found in a merlin territory. (Jack Ashton-Booth)

Diet: Mostly small birds such as pipits, chats, warblers and finches.

Identification: Small, grey and tapered. Often narrower and less rounded than kestrel pellets.

Where found: In and around the nest and favoured perches such as fence or gateposts. Visiting active merlin nests requires a special licence (e.g. Schedule 1) or permit.

Size: 35–88.5mm long and 12–14mm thick.

Notable features: Unlike the kestrel, merlin pellets contain mostly bird feathers.

Hobby

Hobby pellets, revealing insect remains among bat hair. (Collected by Jason Fathers)

A hobby pellet containing insect remains and tiny feathers. (Collected by Jason Fathers)

A hobby pellet containing feathers from a swallow. (Collected by Jason Fathers)

A hobby pellet containing the feathers from a blue tit. (Collected by Jason Fathers)

A hobby pellet containing a mix of insect remains and tiny feathers. (Collected by Jason Fathers)

A hobby pellet containing feathers and a clump of bat hair (bottom right area). (Collected by Jason Fathers)

A hobby pellet containing a mix of bird bones, bird leg skin, claws and small berries from its prey's last meal. (Collected by Jason Fathers)

A hobby pellet from two different angles revealing the matrix of tiny pieces of insect remains. (Collected by Jason Fathers)

Diet: Small to medium-size birds, bats and insects, ranging from moths to dragonflies.

Identification: Variable depending on diet. Smaller and rounder than kestrel or merlin. Those containing just insects' parts are very fragile.

Where found: In and below nest. Visiting active hobby nests requires a special licence (e.g. Schedule 1) or permit.

Size: 20–26mm long and 10–25mm thick (Brown *et al.* 2021).

Notable features: In a selection or pile of pellets some may contain only feathers, some just insect parts and others hair from bats, while some may contain a mix of all three. Larger human or long mammal hair (often from deer or horse) attached to the outside of the pellets will be from the previous nest owners (often crows).

Gyrfalcon

A gyrfalcon pellet containing small feathers and bone fragments, alongside a range of grit and plant material eaten by its prey, most likely a rock ptarmigan, Canadian High Arctic. (Collected by Mike Dilger)

The gyrfalcon is a rare winter visitor from the Arctic and semi-Arctic regions where it summers, although captive/falconry birds may also be found in the wild. In their breeding grounds, prey is mainly dominated by rock ptarmigans and willow grouse, followed by Arctic hares and smaller numbers of songbirds such as snow buntings and Lapland buntings, waterbirds (ducks and waders), ground squirrels and lemmings (Booms & Fuller 2003; Potapov 2011). When in the British Isles gyrfalcons most likely feed on ptarmigans, red grouse (the British subspecies of willow grouse), small songbirds and waterbirds. Visiting active gyrfalcon nests requires a special licence (e.g. Schedule 1) or permit.

Sparrowhawk

A sparrowhawk pellet revealing feathers and a leg, probably from a greenfinch and the tail of another small bird. (Collected by Jason Fathers)

This sparrowhawk pellet contains feathers, foot skin and a metal ring from a greenfinch. (Collected by Jason Fathers)

A sparrowhawk pellet containing small bones (left side) and feathers. (Collected by Jason Fathers)

A sparrowhawk pellet containing a mix of small, fluffy feathers from one or two small birds. (Collected by Jason Fathers)

A sparrowhawk pellet packed full of small bird feathers, with a few orange feathers that may be from a robin. (Collected by Jason Fathers)

A sparrowhawk pellet containing pigeon feathers.

A sparrowhawk pellet containing pigeon feathers and grain eaten by its prey.

Diet: Small to large birds, occasional bats and very young rabbits (first emergence).

Identification: Cylindrical or rounded pellets packed full of small body feathers. Often rounded at each end or tapered at one end. Unlike owls inhabiting the same environment, sparrowhawk pellets mostly lack whole skulls and bony material, the latter of which may get digested or left behind and picked clean (Newton 1986).

Where found: Beneath favourite plucking posts/branches and nests in woodlands, parkland and large gardens. Visiting active sparrowhawk nests may require a special licence (e.g. Schedule 1) or permit.

Size: 25–35mm long and 10–18mm thick (Brown *et al.* 2021).

Notable features: While the wing and tail feathers are often plucked, complete sets of tail feathers and small wings may be swallowed alongside occasional small legs or parts of feet, bills, fragments of bone and the tough sac of the gizzard (Newton 1986). Seeds and insects will be from their prey's stomach or crop contents. Bird identification rings may also be found in pellets. Some pellets resemble those of hobby, although lack the insect remains and found mostly in woodland environments – hobby pellets are mostly found beneath their nest locations, usually in isolated large trees in farmland, heathland and other rural habitats.

Goshawk

A goshawk pellet found after heavy rain revealing the bubbly skin from the sole of a pigeon's foot.

A fresh goshawk pellet containing a pigeon foot, feathers and food from its prey. (Ruth Tingay)

A goshawk pellet, showing its two different sides, revealing small feathers and bone fragments alongside a range of grit, seeds and plant material eaten by its prey.

A goshawk pellet revealing grey squirrel hair and a claw. (Collected by Gareth Jones)

Pellet from a young goshawk (hence its smaller size) revealing the foot of a pigeon and a racing pigeon ring (green). (Collected by Anna Field)

Another pellet from a young goshawk revealing small feathers and grey squirrel hair and claw.

Diet: Range of birds and mammals (especially squirrels and rabbits).

Identification: Variable pellets containing feathers, bones and hair (squirrel or rabbit). Very similar to those of buzzard and red kite. Best identified at known nesting locations and by the presence of moulted feathers.

Where found: Beneath favourite plucking posts/branches and below or in nests. Visiting active goshawk nests requires a special licence (e.g. Schedule 1) or permit.

Size: 30–49mm long and 13–27mm thick.

Notable features: Pellets may contain just feathers or just hair, or a mix of the two. Often includes fragments or bones, whole legs or feet and skulls. Bird identification rings may also be found in the pellets.

Buzzard

A selection of buzzard pellets revealing mostly tightly packed small mammal hair. (Collected by Jason Fathers)

An accumulation of buzzard pellets beneath a nest. (Jack Ashton-Booth)

Diet: Small mammals including rabbit-size prey, birds, amphibians and reptiles, invertebrates (particularly earthworms) and carrion from roadkill or carcasses of animals such as sheep and deer. Many birds brought to the nest may be from road collisions. While small mammals, such as voles, may be eaten whole, flesh is picked off the bones in the case of larger prey.

Identification: Similar in shape to an owl pellet, although contains fewer bones. Buzzards are able to digest their prey more efficiently than other birds of prey (Barton & Houston 1993). Therefore, pellets contain mostly tightly packed hair and/or small feathers (Brown *et al.* 2021).

Where found: Below favourite perches such as tree branches, rocks, telegraph poles and fenceposts and in, below or near nests. Visiting active buzzard nests may require a special licence (e.g. Schedule 1) or permit.

Size: 45–60mm long and 25–30mm thick (Brown *et al.* 2021).

Notable features: Similar to those of other birds of prey such as goshawk and red kite. So, identification best confirmed through known buzzard perches and nest locations and/or the presence of moulted feathers.

Red kite

Diet: A huge range, including birds, mammals, reptiles, amphibians, fish (dying or injured) and invertebrates, both alive and dead. Snatch up smaller animals killed by cars and feed on carcasses of larger animals such as sheep and deer. May also steal prey from a range of other birds of prey, including peregrines, white-tailed eagles and crows (Martin 1992).

Identification: Similar shape and size to buzzards. Largely made of hair and/or feathers, as red kites are very efficient at digesting their prey (Barton & Houston 1993).

Where found: In and below nests, below communal roosts and roosts close to nest sites. Visiting active red kite nests requires a special licence (e.g. Schedule 1) or permit.

Size: Similar to buzzard.

Notable features: Similar to those of buzzard and goshawk. Therefore, identification best confirmed through known red kites perches, communal roosts and nest locations and/or the presence of moulted feathers. Pellets may also contain small rubber castration rings from lambs.

Hen harrier

Hen harrier pellet containing densely packed small mammal hair. (Collected by Jack Ashton-Booth)

Another similar-looking hen harrier pellet full of small mammal hair, from a winter roost. (Collected by Alan McCarthy)

Selection of hen harrier pellets packed full of feathers, many from snipe, a common winter prey species. (Collected by Alan McCarthy)

Hen harrier pellet full of loosely packed snipe feathers. (Collected by Alan McCarthy)

Hen harrier pellet found on the ground near a nest location. (Mike Price)

Hen harrier pellet and poo in a winter roost among rushes. (Mike Price)

Diet: Small mammals, in particular field voles, rabbits and hares, waders, small birds such as meadow pipits, wrens and skylarks and young of larger birds such as grouse. In winter, other gamebirds, thrushes, pigeons and waterbirds such as ducks, geese, rails and waders (especially snipe) may also feature.

Identification: Very similar to other birds of prey, such as owls and marsh harriers, so known locations and moulted feathers are important for identification. Shorter and more rounded compared to short-eared owl, which shares similar breeding habitat.

Where found: Known single-species roosts (outside the breeding season) and nests. At winter roosts, pellet numbers may vary as some may drop through vegetation or get lost with high tides washing them away (Clarke *et al.* 1993). Visiting active hen harrier nests requires a special licence (e.g. Schedule 1) or permit.

Size: 16–82.2mm long and 12–43mm thick (Holt *et al.* 1987).

Notable features: Looks and feels hairier compared to those from a short-eared owl as bones are embedded deeper in the pellet (Holt *et al.* 1987). The poo found near or with pellets can be helpful: hen harriers have white to white/green poo and may have some blackish solid parts, while short-eared owls have buff or creamy poo with a black solid bead (Holt *et al.* 1987).

Marsh harrier

Diet: Small mammals, rabbits and hares, amphibians, eggs, small birds, gamebirds, the adults and young of waterbirds and waders and occasional fish, reptiles and insects.

Identification: Similar to hen harrier. Pellets may sometimes contain only hair or feathers.

Where found: Roosts (outside the breeding season), resting places near nests (up to 200m away (Matos 2015)) and at nests. Also along farm tracks, edges of reedbeds and ditch-sides (Underhill-Day 1985). Pellets may be scarce if they have washed away, dropped through the vegetation or if water levels are too high to access (Clarke *et al.* 1993; Matos 2015). Visiting active marsh harrier nests requires a special licence (e.g. Schedule 1) or permit.

Size: Similar to hen harrier.

Notable features: Best identified from known single-species roosts and at or near nest locations. Moulted feathers at roosts may help with identification. When volunteers are ringing young, marsh harrier pellets are generally not found (Phil Littler pers. comm.). This is probably because marsh harriers keep their nests clean while the nestlings are young and growing. As they get older and begin to explore the nest their pellets are less likely to be removed by their parents, so finding pellets is more likely once the young have fledged and looking at this time also avoids disturbance (Chas & Carter 1925; Colling & Brown 1946).

Golden eagle

A golden eagle pellet revealing the feather remains of a red grouse. When rotated the pellet also reveals a claw from the grouse. (Found by Bernard King, kept at Bristol Museum & Art Gallery)

A dissected golden eagle pellet containing deer hair and small mammal bones, including the jawbone of a stoat or weasel. (Ruth Tingay)

A golden eagle pellet containing hedgehog spines (shown at 60% size). (Ruth Tingay)

A golden eagle pellet containing deer hare and the upper and lower beak of a red grouse (shown at 60% size). (Ruth Tingay)

Diet: Mammals including mice, hedgehogs, pine martens, foxes, badgers and roe deer; birds including small birds such as pipits and larger species such as grouse, crows, ducks, geese and gulls; carrion such as sheep and deer; and occasionally reptiles, amphibians, fish and invertebrates. Diet may contain more rabbits, hares and gamebirds compared to white-tailed eagle (Whitfield *et al.* 2013).

Identification: Thick, substantial pellet although generally smaller than those of a white-tailed eagle. Contains mostly hair or wool and feathers and sometimes teeth or the horny coverings of bird bills or the hoof coverings from sheep or deer. Bones less common and may include occasional legs of birds, bones of rabbits or hares or toe bones from deer or sheep.

Where found: Perches, roosts and nest locations. Visiting active golden eagle nests requires a special licence (e.g. Schedule 1) or permit.

Size: Similar to white-tailed eagle.

Notable features: Similar to white-tailed eagle pellets. Identification best at known locations or using other clues such as moulted feathers. Pellets may also contain small rubber castration rings from lambs.

White-tailed eagle

A white-tailed eagle pellet that has decayed revealing all the fishbones. (Collected by Jason Fathers)

A selection of white-tailed eagle pellets revealing feathers and, in some, the toe skin or webbed toe from large gulls. (Forestry England & Roy Dennis Wildlife Foundation)

A closer view of the webbed toes of a large gull in a white-tailed eagle pellet. (Forestry England & Roy Dennis Wildlife Foundation)

Diet: Fish is preferred when available; birds including ducks, geese, gulls and seabirds from auks to shags; and mammals such as rabbits and hares also feature. Carrion can form an important element of diet, particularly during leaner winter months. In southern England cuttlefish are also eaten (Stephen Egerton-Read pers. comm). Diet often comprises more seabirds, ducks, waders and fish than in golden eagles (Whitfield *et al.* 2013).

Identification: Large unmistakable pellets resembling a dog poo! Dark to light-brown pellets containing a mix of feather, hair or wool, bones and legs/feet. May be lighter in colour depending on food.

Where found: At and under nests, trees used for perching and roosting and feeding locations. Visiting active white-tailed eagle nests requires a special licence (e.g. Schedule 1) or permit.

Size: 90–110mm long and 35–40mm thick, although may be smaller (Brown *et al.* 2021).

Notable features: Similar to golden eagle pellets. Identification best at known locations or using other clues such as moulted feathers. Pellets may also contain small rubber castration rings from lambs.

6 CORVID (CROW FAMILY) PELLETS

Corvid pellets, especially those of raven, carrion and hooded crow, rook and jackdaw, share similar contents at different times of the year, especially grain and ground-dwelling invertebrates. All corvids in Britain and Ireland produce pellets, reflecting their omnivorous diet as they consume a range of foods including seeds, nuts, invertebrates, small birds and mammals, as well as scavenging dead animals such as roadkill. Pellets therefore contain a range of items depending on what the corvid has been feeding on and can be hugely variable. Their pellets may be packed with insect remains and plant material ingested when swallowing beetles and worms. Pellets of ravens, carrion crows and magpies that have been feeding on roadkill may be full of hair or feathers. Bones may be picked clean and not appear in pellets, while oat and wheat husks are common from eating animal feed, ripe crops and feeding in harvested fields. Seeds from fruits such as cherries and rowan berries will be prevalent in summer and autumn.

Many corvid species will roost together in large numbers in woodlands and some, such as the rook, nest colonially. This provides opportunities for finding pellets beneath where they roost or nest. They also spend time watching, observing and resting on fenceposts and gates. So, these are ideal places to look for regurgitated pellets, both on the perches and on the ground beneath. Identification is best ascertained by collecting pellets from known nest or roost locations of particular species, or haunts where certain species are confirmed.

The photos for different species in this chapter show just some of the varieties of corvid pellets and those represented for rook may also be similar to those for crow and jackdaw at certain times of the year. Subtleties are discussed below that may help distinguish between species – although it is not fool-proof!

Note: All images of pellets and pellet contents are shown life size unless a scale bar or a note in the photo caption indicates otherwise. Outdoor, 'in-the-field' images are not to scale.

Whodunnit? These are from different, unidentified corvids and show the variety of pellets produced.

A small, narrow corvid pellet found on a disused airfield.

A small corvid pellet revealing a dragonfly wing on the left side.

A corvid pellet full of click, ground and small dung beetles.

A pellet revealing grey squirrel hair, probably from roadkill.

A small, narrow pellet, probably from a magpie, left on a field gate.

A weathered magpie or crow pellet containing a fruit stone.

A raven or carrion crow pellet, 70mm long, found on a stone wall. (Bob Cowley)

A raven or carrion crow pellet found with other similar pellets on sand dunes. The tiny snail shells are from land snails, known as pointed snails, and are only found on sand dunes and chalk downland up to a few miles inland from the sea. The pellet is very fragile as it made up of loose sand and grit granules. (Collected by Rhian and Ben Rowson)

Pointed snail shells and remains of a woodlouse in a corvid pellet. (Collected by Rhian and Ben Rowson)

Raven

A raven pellet revealing a mix of hair, plant material and bones. (Collected by Richard Clarke)

A raven pellet revealing insect egg cases or plant capsules – microscopic examination was inconclusive. (Collected by Richard Clarke)

A raven pellet revealing a matrix of eggshell and plant material. (Collected by Jason Fathers)

A freshly regurgitated raven pellet full of wild fruits such as rowan berries and seeds. (Jack Ashton-Booth)

A raven pellet containing sheep's wool. (Collected by Julian Driver)

Diet: A huge range of food including carrion (especially dead sheep and lambs in the uplands), small mammals including moles, invertebrates from the shoreline (such as mussels and crabs) and grassland/moorland habitats, birds and seaweed.

Identification: Large cylindrical pellets. In the uplands often dominated by sheep's wool (and other livestock) and the hair of rabbits, hares or smaller mammals such as voles. On lower ground, pellets are more similar to those of other crows, containing small bones, some plant material and the hard parts of insects, or to those of birds of prey such as buzzards, especially if they have been feeding on small mammals (Hardey *et al.* 2013).

Where found: Nests and below communal roosts, on stone walls and top of fenceposts.

Size: 50–80mm long and 20–30mm thick (Ratcliffe 1997).

Notable features: May contain bones, racing pigeon rings and small rubber castration rings from lambs (which are also found in kite and eagle pellets) (Hardey *et al.* 2013). Pellets give off a distinctive musty smell – similar to the smell of wet dog hair – characteristic of corvids.

Rook

Rook pellets found on the ground in a field close to a rook colony.

A rook pellet after recent rain showing how quickly they disintegrate.

Rook pellets collected beneath a nesting colony. These contain wheat grain kernels probably eaten from nearby gamebird feeders or stubble fields.

Diet: Grains, fruits and invertebrates such as earthworms, beetles and fly larvae (feeding on more invertebrates that live underground compared to jackdaws), acorns and fruit.

Identification: Fibrous, cylindrical and often tapered pellets.

Where found: Beneath nest colonies, fenceposts and on or below farm gates.

Size: 30–40mm long and 10–15mm thick (Brown *et al.* 2021).

Notable features: Variable depending on diet. Easy disintegrates during rainy weather. A greater quantity of soft-bodied soil invertebrate larvae in the diet means invertebrate remains in pellets are less common than in other corvids (Losito & Cowley 2020).

Carrion crow

Carrion crows gathering on a stone wall by sheep fields, an ideal place to look for their regurgitated pellets. (Bob Cowley)

A carrion crow pellet full of grain husks, small mammal hair and a tiny stone. Found along a mown footpath through a tall grassy meadow.

Diet: Grain, invertebrates such as earthworms and beetles (a mix of those living underground and above ground). An intermediate feeder to jackdaw and rook, with pellets containing a bit of what each of these two species feeds on (Lockie 1956).

Identification: Fibrous, cylindrical and often tapered pellets.

Where found: At or near nests, favourite regular perches below trees or on stone walls and fenceposts and below communal roosts. Also in fields and along beaches or the shoreline where they feed.

Size: 30–70mm long and 10–20mm thick (Brown *et al.* 2021).

Notable features: Generally larger than rook pellets, denser and richer in dung beetle remains (Losito & Cowley 2020).

Hooded crow

Diet: Grain (winter), invertebrates (summer), fruit, carrion, fish, crustaceans, molluscs such as snails and bivalves and small mammals. Occasionally small birds and eggs.

Identification: Fibrous, cylindrical and often tapered pellets.

Where found: At or near nests, favourite regular perches below trees or on stone walls and fenceposts and below communal roosts. May also be found in fields and along beaches or the shoreline where they may feed.

Size: 30–70mm long and 10–20mm thick (Brown *et al.* 2021).

Notable features: Very similar to and resembling those of carrion crows and other corvids.

Jackdaw

A jackdaw pellet found in an orchard (Paul F. Whitehead). In this one pellet were found the following: *Amara aenea*, *Harpalus rufipes*, *Poecilus cupreus*, *Pterostichus madidus*, *Philonthus cognatus* and *Agriotes lineatus*. Mammal dung specialists: *Onthophagus coenobita*, *Agrilinus ater*, *Acrossus rufipes* and an *Aphodius* sp.; the arboreal longhorn beetle *Anaglyptus mysticus*, and the weevil *Liophloeus tessulatus*. Other insects eaten include the arboreal hawthorn shieldbug *Acanthosoma haemorrhoidale*, meadow grasshopper *Pseudochorthippus parallelus* and the common wasp *Vespula vulgaris* (Whitehead 2022).

Selection of pellets containing mostly common dor beetle remains. Most likely from jackdaw, although choughs were sighted around the area during the same period. (Collected by Nigel Massen)

Diet: Weed seeds, snails, invertebrates living overground such as ants and beetles (and smaller than those taken by rooks), moth larvae, acorns, fruit and grain.

Identification: Fibrous or textured, cylindrical and often tapered pellets.

Where found: Near nests and communal roosts.

Size: 25–30mm long and 10–15mm thick (Brown *et al.* 2021) and similar in size to pellets of kestrels and little owls.

Notable features: Similar in shape and size to those of other corvids, so identification best confirmed from known nest areas and roosts.

Magpie

Magpie pellet, 45mm in length, containing soil, fruit stones, grit and parts of a crab.

Magpie pellet, 32mm in length, containing soil, fruit stones, woodlice and vegetable matter.

Diet: Hugely variable and changes throughout the seasons, ranging from fruits, grain and vegetation in the autumn/winter to ground and underground inverte-brates (including beetles, ants, caterpillars, spiders, larvae of craneflies (leather-jackets) and earthworms) during the summer months. To a lesser degree, magpies also feed on small mammals and birds and other invertebrates, some of which may be scavenged as carrion (Tatner 1983; Birkhead 1991).

Identification: Cylindrical or rounded shape; may have some tapering at one end. Best collected from known magpie sources to be sure of identification.

Where found: Beneath nests and communal roosts.

Size: 35–45mm long and 10–20mm thick (Brown *et al.* 2021).

Notable features: Fragile and easily disintegrate during rain; best collected after dry nights (Birkhead & Clarkson 1985).

Chough

Five chough pellets collected as part of diet studies in Spain in the 1980s (Solar & Solar 1993). (Zoology Department of Granada University)

Diet: Invertebrates such as dung and chafer beetles (adults and larvae), ants, earwigs, earthworms, fly larvae and grasshoppers/crickets; wild seeds and cultivated grain.

Identification: Similar to other corvid pellets, in particular jackdaw, from birds that have been feeding on invertebrates or grains.

Where found: Feeding and roosting locations (if accessible).

Size: 19–24mm long and 11–13mm thick.

Notable features: If the diet comprises mostly soft-bodied soil invertebrates it is possible that mouthparts and head capsules pass through the choughs' digestive system and come out in their poo rather than as pellets, although they may be entirely undetectable (McKay 1996). Grit may feature in pellets and appears to be associated with feeding on cultivated grain (Solar & Solar 1993).

Jay

Jay pellets are hard to find due to the shy nature of the species, which tends to inhabit a denser woodland environment compared to other corvids, and where their roosts are generally difficult to find. However, occasionally they may use roosts that are more obvious – for example, in the city of Kharkiv, Ukraine, jays mostly use Norway spruce and poplar trees with pre-roost groups varying between four and 15 birds (Bresgunova 2014). Where roosts are known, pellets are more likely to be encountered beneath. In Italy, jay pellets were only found in the Mediterranean scrub maquis and comprised the remains of cicadas, scarab beetles, longhorn beetles, grape seeds and fruit from blackthorn and brambles (Rolando 1998).

7 GULL, TERN AND SKUA PELLETS

Seabirds, including gulls, terns and skuas, eat a huge variety of food from the sea, ranging from fish of many different sizes through to small crustaceans such as crabs. Many marine invertebrates are hard-shelled and, even those that are soft-bodied, such as worms, have hard parts such as jaws – all of which may be regurgitated in pellets. Fish feature highly in the diets of many seabirds and their distinctive vertebrae, ear bones (otoliths) and other parts are important clues to finding out what the birds have been eating.

Larger gulls and skuas, in particular the great black-backed gull and great skua, will swallow animals whole – including small rabbits and seabirds such as puffins, petrels and young auks. This means much of their skeletal and feather or hair remains are regurgitated in pellets.

Note: **All images of pellets and pellet contents are shown life size unless a scale bar or a note in the photo caption indicates otherwise. Outdoor, 'in-the-field' images are not to scale.**

Terns

A pellet, 10mm diameter, from a sooty tern containing fishbones and a fish otolith (white spot on the right-hand side), Trindade Island. (Dr Leandro Bugoni, Universidade Federal do Rio Grande (FURG))

Terns produce rounded pellets which can be collected at roosting locations adjacent to nesting colonies, in nesting colonies when opportunities are afforded by the likes of nest-recording and ringing of young (to minimise any disturbance), and at non-breeding roosts. Single-species roosts are usually targeted. While fish dominate the diet, insects such as dragonflies and diving beetles, and crustaceans, may also be important depending on the tern species and the time of the year. Sandwich tern pellets measure 18–25mm in length and 11–15mm in diameter, while those for little terns may be 13–16mm in length and 9–11mm in diameter (Below 1979). For more details on tern diet and using pellets to identify their prey, see Mauco *et al.* 2001, Granadeiro *et al.* 2002, Bugoni & Vooren 2004 and Catry *et al.* 2006. Visiting active tern colonies may require a special licence (e.g. Schedule 1) or permit.

Herring and lesser black-backed gulls

An urban-dwelling herring or lesser black-backed gull pellet regurgitated onto the roof parapet where the bird had been resting.

These herring or lesser black-backed gull pellets have been collected from urban rooftops. They are variable in shape and size, although frequently contain plant material such as grasses which the birds swallow incidentally with their food. These pellets contain a range of litter including a plastic bag, foil, chicken bones and glass.

A herring or lesser black-backed gull pellet that has disintegrated revealing numerous tiny plastic pieces including nurdles.

A pellet from a herring gull on a coastal island revealing the remains of a crab and fish vertebrae. (Jack Ashton-Booth)

A pellet from a coastal-dwelling herring gull found on grassy ground. It reveals mostly fish vertebrae and fine downy feathers. (Alan Rowland)

Diet: Coastal gulls forage and scavenge along the coastline and over the sea. Lesser black-backed gulls may also surface-feed at sea on crustaceans, such as swimming crabs, while herring gulls also consume bivalves, such as mussels, often dropping them from a height to smash them open. In the past fishery discards, now banned in the UK, were important sources of fish. These birds will also venture inland to feed on invertebrates and small mammals in agricultural fields and pick up littered food in seaside towns and villages. Thousands of urban-dwelling gulls forage in cities and towns and commute out to fields to feed on soil invertebrates. Both herring gulls and lesser black-backed gulls will feed on a range of discarded takeaway food, food snatched from people (from ice creams to pasties) and refuse. Natural food includes invertebrates such as earthworms, ants, beetles and crustaceans, freshwater fish, small mammals, birds, roadkill and carrion washed up on the shoreline.

Identification: Generally round or globular in shape, fibrous and often tapered at one end. Variable depending on diet. Pellets frequently contain vegetation probably swallowed when feeding on earthworms.

Where found: Found among nesting colonies and at resting/loafing locations on rocky shores, playing fields and along breakwaters and jetties. On buildings, found on roofs and parapets where the gulls nest. Those of lesser black-backed gull may be easier to find – they nest on flat open roofs of buildings while the herring gull more often nests on chimneystacks and other difficult-to-access locations that replicate cliffs. Pellets may also be found at ground level, for example on playing fields and green spaces where they rest up during the day. In more traditional coastal areas, pellets can be found around nesting colonies, beaches where they are resting and favourite perches and lookouts.

Size: 25–50mm long and 15–20mm thick (Brown *et al.* 2021). See 'Fish' on page 230 for reference to a gull pellet analysis guide identifying marine contents.

Notable features: In urban locations pellets will contain a mix of plant material such as grasses, soil, particles of plastic and glass, grit, plastic bags (and even condoms) and anything that is easily swallowed. Pellets from coastal gulls may also contain a mix of these things, although fishbones, crab legs and carapaces, small (including voles) and medium-size mammals (such as rabbits), eggshells and seabirds will also appear. Gulls will even regurgitate their pellets in flight!

Great black-backed gull

A great black-backed gull pellet revealing the hair and small bones of a rabbit.

A great black-backed gull pellet revealing the legs and feathers of a puffin.

A great black-backed gull pellet containing the remains of a rabbit including connected leg bones. (Ellie Hilton)

A disintegrating great black-backed gull pellet revealing rabbit hair and bones. (Ellie Hilton)

Diet: Marine fish, seabirds (from puffins to petrels) and their young (terns and gulls), marine invertebrates including crabs, shellfish and crustaceans, carrion and waste rubbish/refuse. As with herring and lesser black-backed gulls, fishery discards used to be important sources of fish. They also harass other seabirds until the victims regurgitate their fish (kleptoparasitism).

Identification: Larger and bulkier than herring or lesser black-backed gulls, often containing larger prey remains such as small rabbits and seabirds such as puffins, shearwaters and petrels.

Where found: Near nesting areas and at resting/loafing locations on rocky shores and jetties.

Size: 35–55mm long and 15–25mm thick (Brown *et al.* 2021).

Notable features: Pellets containing feather and hair are more tightly packed and hold their shape for longer without disintegrating. Those containing fishbones, including otoliths, are looser and more fragile (Buckley 1990).

Great skua

A great skua pellet revealing the bill and feathers from a puffin.

A great skua pellet revealing the foot and bones from a puffin.

A whole puffin leg and foot that came out of a great skua pellet.

58mm

54mm

A great skua pellet containing a Leach's petrel, swallowed whole, digested and regurgitated with the wings and legs still intact.

Diet: Marine fish, seabirds and their young, invertebrates such as goose-barnacles, eggs, mammals including small rabbits, and carrion. Fishery discards were also important before being banned.

Identification: Variable in size, colour and texture depending on prey. Large and irregular shape, often containing the remains of whole or parts of seabirds or rabbits. Pellets may also be full of fishbones. Many fishbones do get digested or damaged, though, making their identification to species more difficult (Votier *et al.* 2003).

Where found: Close to nests and washing/resting locations. Visiting active great skua nests may require a special licence (e.g. Schedule 1) or permit.

Size: Similar in size to those of great black-backed gulls.

Notable features: As with other birds, the pellets are full of feathers and/or hair when they have been feeding on birds and mammals (either hunted or scavenged). When feeding on eggs, the pellets may contain fragments of eggshell (often reduced to powder). Great skua chicks apparently do not produce pellets (Furness 1987), probably because they are fed small whole fish that are easily digested or just the flesh of birds and larger fish. However, young in captivity fed on a range of birds and fish will produce pellets (Votier *et al.* 2001). Pellets are likely to show a bias towards bird prey rather than fish prey, the latter of which is more easily digested compared to feathers. They may also over-represent the number of petrels and young birds, which are swallowed whole, compared to larger birds, which are plucked or torn apart and whose feathers are less likely to appear in the pellets as a result (Votier *et al.* 2001).

If great skuas have been feeding on marine life their pellets can be variable depending on what fish or invertebrate species they have been consuming, as outlined below (Votier *et al.* 2003):

- A diet of sandeels produces small pellets full of tiny grey bones (especially vertebrae) and tiny grey scales. They also contain many small, oval white otoliths.
- If great skuas have been feeding on a range of whitefish such as haddock or whiting their pellets comprise larger, white fishbones and a few larger white otoliths.
- Pellets containing herring or mackerel are greyish-white with thin fishbones, many large, greyish-white scales and usually no otoliths.
- If great skuas have been feeding on goose-barnacles their pellets are full of broken fragments of white plates that resemble shell fragments of smooth bivalve molluscs.

Arctic skua

Diet: Fish, berries, birds, small mammals, invertebrates (ranging from insects to crustaceans) and eggs. Some food, especially fish and marine invertebrates, is obtained through kleptoparasitism – chasing other seabirds until they regurgitate their food.

Identification: Small pellets.

Where found: Territorial mounds.

Size: No data.

Notable features: Pellets are found in small numbers and can be uncommon or rare. The nature of their diet (soft foods and smaller birds compared to great skua) sees food passing through the digestive tract and therefore examining the poo of this species may offer more reliable information on diet (Andersson & Götmark 1980; Smith & Jones 2006).

8 OTHER SEABIRD AND WATERBIRD PELLETS

Seabirds and waterbirds fill many different niches across watery environments including fast-flowing and slow-flowing rivers, lakes and ponds, marshland and bogs, beaches and saltmarsh, rockpools and the deeper, open sea. Finding and studying seabird and waterbird pellets provides an opportunity to learn more about their rich and varied diets across many different species. The selection of species featured in this chapter reflects the range of pellets produced by different species exploiting these many habitats. They feed on prey that may include fish and small mammals in the case of the grey heron, frogs and grasshoppers for white storks and small fish for the kingfisher.

Wading birds or shorebirds come in all sorts of shapes and sizes, feeding on a variety of prey and suitably adapted to particular environments with their different sizes and shapes of bills. Our largest wader, the curlew, which feeds on larger marine worms and shellfish can also consume some of the smallest prey such as mud snails. Cormorants and shags produce mucus-rich pellets that turn rock-hard once dry, providing robust material full of fishbones and marine invertebrate remains that can be easily handled and examined. Like the kingfisher, the dipper prefers streams, brooks and rivers where it swims to find underwater invertebrates among pebbles and stones, leaving pellets on regularly used stones highlighted by its tell-tale white poo. Grebes also produce pellets, probably daily, although there have been very few pellets found or observations of pellets being ejected. They are thought to digest most fishbones and the pellets largely comprise vegetable matter and often, although not always, feathers (see 'Feathers', p. 216). Some fishbones and invertebrate remains may also be encountered. Unless a grebe is in captivity, pellets are regurgitated in the water and so cannot be collected (Storer 1961; Kop 1972; O'Donnell 1982; Wiersma *et al.* 1995).

For ducks, most foods are digested or pass through the digestive system and come out as poo, including shellfish remains, where the shells are swallowed and broken down in the gizzard (Trewin & Welsh 1976). However, ducks may regurgitate parts of their food, such as seeds, as loose pellets or larger boluses if they have been over-eating and gorging, or if the indigestible parts of their food are too large to be passed through their digestive system (Tarshis & Rattner 1982; Kleyheeg & van Leeuwen 2015). For example, in the next photo, this hooded merganser was overfeeding and gorging on crayfish (which are usually digested within a few hours). It can be seen regurgitating what it could not quite manage to digest!

Note: All images of pellets and pellet contents are shown life size unless a scale bar or a note in the photo caption indicates otherwise. Outdoor, 'in-the-field' images are not to scale.

A hooded merganser regurgitating partly digested crayfish, Ontario, Canada. (Melanie Howarth – Flickr profile: Slow Turning)

Procellariiforms – seabirds such as albatrosses, fulmars, shearwaters and petrels – apparently rarely regurgitate pellets, meaning that any hard, undigested remains (alongside pieces of plastic) remain in their stomachs for a long period (Imber 1973, Barrett *et al.* 2007). More often these birds regurgitate semi-digested or undigested food when handled, contributing towards diet studies (Barrett *et al.* 2007). However, one group, the *Pterodroma* petrels – which are seldom encountered in Britain and Ireland – appear to produce pellets more reliably, containing squid beaks and feathers (swallowed during preening) (Imber 1973, Leal *et al.* 2017).

A pellet, 22mm diameter, from a Trindade petrel containing fishbones, squid beaks (dark reddish brown) and feathers, probably from a chick that has been preening and eating them during the process, Trindade Island. (Dr Leandro Bugoni, Universidade Federal do Rio Grande (FURG))

Grey heron

A selection of grey heron pellets revealing the soft, pad-like pellets containing mostly small mammal hair and a little bone. (Collected by Kane Brides and studied by María José Navarro-Ramos)

A selection of grey heron pellets found beneath a cypress tree. (Collected by Kate MacRae)

A pellet regurgitated by a grey heron chick. It is more varied than the others shown, revealing plant matter and fish scales. (Collected by Barry Trevis)

Diet: Fish, invertebrates such as crustaceans and beetles; amphibians and reptiles; small mammals, including field voles, water voles and moles; birds such as little grebes and ducklings.

Identification: Large cylindrical dark brown pellets, often resembling those of barn owls although softer, less bony and looser around the edges.

Where found: Beneath heron colonies, in nests and below favourite resting spots. Grey herons can be very sensitive to disturbance and visits should be kept to a minimum, ideally timed with nest-recording checks or ringing of chicks. Visiting active grey heron colonies may require a special licence (e.g. Schedule 1) or permit.

Size: 15–90mm long and 25–40mm thick (Giles 1981; Brown *et al.* 2021).

Notable features: Away from nesting colonies, where identification is easier, grey heron pellets may resemble those of owls, particularly when they have been eating small mammals such as voles. Unlike owls, herons digest many of the bones. Therefore, pellets tend to be soft hairy packages (especially once dry) with some bones, particularly teeth and skulls. Grey herons swallow vegetable matter, which is thought to help form pellets, and in turn helps to disperse seeds of plants and eggs of invertebrates (Lowe 1954; Navarro-Ramos *et al.* 2022).

Great white egret and bittern

A great white egret regurgitating a pellet containing the hard parts of crayfish, Sequoyah National Wildlife Refuge, Oklahoma, USA. (Steve Creek, stevecreek.com)

The great white egret is a rare and relatively new breeding species in Britain and great care is required when approaching nests for research. Pellets are found beside and below nests, depending on whether they are high in trees or at the water level in reedbeds. If among reedbeds, the pellets may easily get lost in the water. The pellets of nestlings are those most often found. A mean length of 35mm and width of 22.8mm was found in a study by Figueroa & Stappung (2003), while a mean length of 42mm was found in a study by Pretelli *et al.* (2012). Their diet varies depending on the location and mostly includes fish and crustaceans (crayfish), alongside smaller proportions of insects (including dragonflies) and the occasional rodent or young waterbird or wader. As with grey herons, the bones of small mammals are for the most part fully digested (Figueroa & Stappung 2003; Pretelli *et al.* 2012).

Bitterns are also rare in Britain and Ireland, although their population is increasing across suitable wetland habitats. They require due care and minimal disturbance at the nest, where pellets can be found. Visiting active bittern nests requires a special licence (e.g. Schedule 1) or permit. Prey is dominated by fish, although amphibians may also be important, especially for nestlings. Crustaceans, small mammals and aquatic invertebrates are also taken (Poulin *et al.* 2007; Brown *et al.* 2012).

A pellet containing the remains of a crayfish found along the rocky edge of a pond; it is most likely from a species of egret or heron, Villanueva de Valrojo, northern Spain.

Part of a crayfish appendage removed from a pellet, Villanueva de Valrojo, northern Spain.

Little egret

Little egrets often nest among grey herons and, in other parts of Europe, with other species such as cattle egrets. They will also nest as their own single-species colony and may roost at the same location outside of the breeding season. Like grey herons, little egrets can be very sensitive to disturbance and visits should be kept to a minimum, ideally timed with nest-recording checks or ringing of chicks. Visiting active little egret colonies may require a special licence (e.g. Schedule 1) or permit. Pellets are found beside or below nests and roosting/perching areas (Buatip *et al.* 2014). Their diet is mostly fish, although crayfish may form an important proportion in some populations. Other prey includes beetles, grasshoppers and crickets, and occasionally small mammals and amphibians (Salazar *et al.* 2005).

White stork

A wild white stork pellet produced by a bird that was being supplementarily fed. Alongside undigested parts of one-day-old chicks (domestic chicken) – the sprats are largely digested – there are also grasshoppers and beetles that have been foraged in the grassland. (Dawne Davis, Knepp)

A selection of white stork pellets produced by outdoor captive storks fed on one-day-old chicks (domestic chicken) and sprats. The pellets contain mostly feathers and some plant material. (Collected by Dawne Davis, Knepp)

Diet: Amphibians, reptiles, fish, invertebrates (such as earthworms, grasshoppers and crickets) and small birds.

Identification: Rounded (less cylindrical compared to grey heron) pellets containing a fibrous mix of vegetation matter and soil alongside hair, feathers and invertebrate body parts (such as those of grasshoppers and beetles).

Where found: In and beneath nests and favourite resting spots.

Size: 40–65mm long and 25–35mm thick (Brown *et al.* 2021)

Notable features: Slightly smaller than those of grey heron, white stork pellets reflect a diet of foraging among wildflowers and grassy meadows and wetland habitats, picking up vegetation and soil as they catch earthworms, frogs, invertebrates and small mammals. As with grey herons, storks digest most bones. Fresh pellets have a sickly sweet and unpleasant smell (Olsen 2013).

Cormorant and shag

Cormorant and shag pellets at the coast can be found at low tide (and avoiding disturbing any nesting birds) on favourite rocky ledges, buoys and floating walkways.

The environment in which shag and cormorant pellets may be encountered – this is the base of a nineteenth-century coastal fortification tower. (Ian Buxton)

An accumulation of shag and cormorant pellets in a crevice above which the birds rest. (Ian Buxton)

A fresh shag or cormorant pellet, with wet mucus, revealing the lower pharyngeal jawbone and teeth of a wrasse, probably ballan wrasse. (Ian Buxton)

A mix of cormorant and (mostly) shag pellets containing a range of fishbones, shells and parts of crabs. Many pellets also contain small smooth stones. Some are embedded within the matrix of the pellet, while others seem stuck to the outer mucus coating. (Collected by Ian Buxton)

A cormorant pellet revealing the orange fins from perch, Lake IJsselmeer area, Netherlands. (Stef van Rijn)

A cormorant pellet containing a mix of fishbones and with moulted feathers stuck to it, Lake IJsselmeer area, Netherlands. (Stef van Rijn)

A cormorant pellet revealing fishbones, including pharyngeal bones (from cyprinid fish), Lake IJsselmeer area, Netherlands. (Stef van Rijn)

A cormorant pellet covered in otoliths (fish ear bones), Lake IJsselmeer area, Netherlands. (Stef van Rijn)

Diet: Fish and marine invertebrates.

Identification: Pale creamy-yellow or brown, grey or whitish knobbly pellets packed full of fishbones (including small jawbones with teeth), shells, scales, hard parts of crustaceans, polychaete worm jaws and small pebbles. Some of these may be secondarily consumed from fish that had these in their stomachs. Pellets are covered in a distinctive mucus known as the cuticula gastris or koilin. This originates from the koilin membranes that protect the lining of the gizzard.

Where found: Nesting colonies, beneath tree roosts, sandflats and on jetties, buoys, beacons and rocky ledges where the birds rest after feeding. Visiting active cormorant or shag colonies may require a special licence (e.g. Schedule 1) or permit.

Size: 10–70mm long and 10–50mm thick (van Eerden & van Rijn 2022).

Notable features: Pellets almost exclusively contain parts of fish and marine invertebrates – unlike those of gulls which contain more vegetation, plastic and bones from birds and small mammals. When fresh, cormorant or shag pellets are slimy. However, when dry they turn hard like glue making them very easy to handle and examine. Plastic litter may be found in some pellets (Acampora *et al.* 2017). Pellets may contain small stones or pebbles; their presence may be incidental – ingested while grabbing prey from the sea floor or being from the stomachs of prey – or potentially swallowed intentionally to help get rid of stomach parasites such as nematode worms (Robinson *et al.* 2008). The size of the pellet is determined by the type and size of fish eaten.

TABLE 1: Six types of pellets can be identified depending on what cormorants have been feeding on, as outlined by van Eerden & van Rijn (2022) based on pellets collected in the Netherlands.

| TYPE | SIZE RANGE (MM) | | DESCRIPTION |
	LENGTH	WIDTH	
Whitish	40–70	30–50	Whitish with lots of fishbones, fin rays and vertebrae loosely packed; thin mucous lining.
Grey large	40–60	30–40	Greyish with lots of bony structures, fin rays, scales and vertebrae loosely packed.
Grey small	20–35	15–20	Greyish with fin rays, scales and vertebrae loosely packed.
Yellow large	40–50	20–30	Yellowish or brownish with partly visible fishbone structures; yellow-brown soft content.
Yellow small	20–35	15–20	Yellowish or brownish sometimes slimy, irregular structure; yellow-brown soft content.
Bloody	10–30	10–20	Yellowish or brownish, with partial or complete bloody lineation.
Loose wrap	10–30	10–20	Slimy or jelly soft, whitish or yellowish lineation with no clearly visible fish remains.

Kingfisher

A kingfisher regurgitating a pellet containing fishbones and parts of small invertebrates. The nictitating membrane can be seen across the bird's eye. (Roger Eads)

A kingfisher pellet that has fallen onto leaves. More often they fall into the water. (Robert Fuller, robertfuller.com)

An adult kingfisher holding a freshly regurgitated pellet in its bill. (Robert Fuller, robertfuller.com)

An adult kingfisher picking at a fresh pellet. These delicate pellets are broken up easily by the adult and added to the nest material surrounding them. (Robert Fuller, robertfuller.com)

A young kingfisher regurgitating its own pellet. The young are sat on a nest of fishbones, all made from disintegrated pellets. (Robert Fuller, robertfuller.com)

Diet: Young coarse fish such as dace, roach, perch, chub, brown trout and adult minnow, three-spined stickleback, bullhead and stone loach. Non-fish prey is rare and is likely to be where kingfishers are mistaking other animals for fish, although they will sometimes chase after insects: these may include adult dragonflies, diving beetles and crayfish. Very rarely newts are caught. On the coast, their diet may also include prawns and gobies.

Identification: Small, oval or round white pellets packed full of tiny fishbones and scales.

Where found: Nest locations and favourite perches (although often pellets fall into the water). Visiting active kingfisher nests requires a special licence (e.g. Schedule 1) or permit.

Size: 10–40mm long and 5–15mm thick (Brown *et al.* 2021).

Notable features: Kingfisher pellets do not stay whole for long. They are delicate and fragile. In the nest burrow, pellets get trodden into the ground forming a layer of small fishbones. Out of the breeding season they may help indicate a recently used burrow. See 'Fish' on page 230 for references on identifying small fishbones.

Wading birds

A redshank regurgitating a pellet, United Arab Emirates. (Mike Barth – Flickr profile: Mike Barth Photography)

Diet: Wading birds have a varied diet depending on the niche that they fill with their different sizes and shapes of bill, as well as the season and their preferred habitat. Foods may include marine invertebrates (including mud snails, crabs, bivalves and shrimps); earthworms; small fish; adult insects and their larvae including beetles, craneflies and midges; and small seeds. Larger wading birds, such as curlews, may occasionally eat small mammals.

Identification: Often cylindrical, granular, elongated pellets packed full of tiny snail shells, larger shell fragments, parts of other invertebrates such as crabs, isopods and polychaete worms (jaws), sand granules and soil. Very delicate with bits easily falling away when dry. If feeding on crustaceans, such as freshwater shrimps, pellets are softer and smooth with a cinnamon-orange hue. Fresh intact redshank pellets are oval and jellybean-like in shape, slightly curved with ridge along their outer curve (Goss-Custard & Jones 1976). Oystercatchers feeding on shellfish, such as mussels, produce pellets containing shell fragments and opercula – the hardened structures that form trapdoors at the entrance of marine snails such as common periwinkles (Trewin & Welsh 1976). Pellets of sanderlings are mostly made up of sand and fragments of crustaceans and snail shells (Stuart & Stuart 2013).

Where found: Along the foreshore where wading birds rest and at high-tide roosts. When wading birds are resting at high tide they are preening, sleeping and regurgitating pellets. However, such places are important refuges and should never be disturbed at high tide or several hours either side (low tide is best and only if

A greenshank regurgitating a pellet. The cinnamon-orange colour is where the remains of their food (crustaceans) have changed colour – just like prawns go pink when cooked. (Dave Newbold)

accessible). If pellets are collected from roosts at low tide they can often be identified by the adjacent footprints (Goss-Custard & Jones 1976).

Size: Redshank pellets are 20mm long and 10mm thick; curlew pellets may be up to 60mm long (Goss-Custard & Jones 1976); sanderlings measure 9–14mm long and 6–8mm thick (Below 1979); avocet pellets are on average 10mm thick (Stuart & Stuart 2013); and green sandpiper pellets measure 17–20mm long and 7.5–9mm thick.

Notable features: Pellets may contain the chaetae and jaws of marine worms, shells from small bivalves and gastropods and parts of crustaceans, in particular crabs. When fresh they may have the odour of shrimps. Seeds may also feature in pellets of certain waders, potentially contributing towards the dispersal of plants (Lovas-Kiss *et al.* 2019).

Green sandpiper pellets from birds regularly using watercress beds during the winter. The birds busily feed on freshwater shrimps and are observed regurgitating pellets every 15–20 minutes (Holt & Warrington 1996). The pellets come out cinnamon-orange and fade to a ground colour as they dry. These pellets are very delicate and deteriorate quickly in wet weather. (Collected by Barry Trevis)

Dipper

A dipper stands next to a freshly regurgitated pellet. (Stephen Shaw – Flickr profile: Scuba`Steve`)

Dipper pellets, revealing their fibrous texture made up of small undigested parts of underwater invertebrates such as caddisfly and mayfly larvae. (Collected by Mervyn Greening)

Diet: Aquatic insects, in particular their larval forms living in fast-flowing freshwater streams, rivers and pools.

Identification: Small cylindrical fibrous pellets resembling squirrel or guinea pig poo in shape and size! Some may be shorter and more rounded. Dark brown, when feeding on caddisfly and mayfly larvae; cinnamon-orange or pink when fresh and feeding on freshwater shrimps. Lighten to grey-brown when dry. Pellet size, shape and their location where found (i.e. running-water environments) help with identification.

Where found: On mid-river rocks and ledges when medium height water levels have remained steady (heavy rainfall and increased water levels will wash them away).

Size: 12–16mm long and 7.5–8mm thick.

Notable features: Contain tiny fragments of freshwater invertebrates with some standing out as black, shiny particles depending on prey. Even when dry and old may have the odour of shrimps. Pellets may also contain tiny plastic particles (D'Souza *et al.* 2020).

9 GARDEN BIRD PELLETS

Many species of familiar garden or woodland birds produce pellets, including robins, blackbirds, song thrushes, dunnocks, woodpigeons and occasionally starlings (with fruit stones). If songbirds feed on invertebrates with hard exoskeletons then the wing, cases, legs, mouthparts and body segments all need to be regurgitated as a pellet. Ants and other invertebrates, such as spiders, often appear whole, although some birds – such as the green woodpecker – will pass the ants through into their poo. They form distinctive cylindrical poos coated in white uric acid and packed full of ant exoskeletons. Green woodpeckers, along with wrynecks, will also produce pellets full of ants. Other woodpeckers, such as the great spotted, may produce pellets containing the indigestible parts of pine cone seeds. Where garden or woodland birds have been feeding on small fruits with seeds, such as blackberries, raspberries and cherries, or seeds with hard outer coatings, the undigestible parts may also be regurgitated as a pellet. Grit, wheat chaff, seed husks and soil may also be incorporated into a pellet – for instance, woodpigeons may produce pellets containing beechmast or husks of barley and oats (Tucker 1944).

Where to find pellets in your garden

Look for pellets in areas where birds are often feeding, resting or nesting. Small bird pellets are easier to find on concrete or paved areas, while larger pellets from magpies and crows may be found on lawns. Fenceposts and gates are also excellent places to seek pellets. Birds, such as robins, nesting in outbuildings and porches may also leave small (10mm long) pellets – round or oval-shaped – beneath or near their nests. They are very delicate and fall apart easily.

Note: All images of pellets and pellet contents are shown life size unless a scale bar or a note in the photo caption indicates otherwise. Outdoor, 'in-the-field' images are not to scale.

Robin and blackbird-size birds

A robin regurgitating a pellet. (Robin Morrison)

A blackbird regurgitating a pellet. (Peter Staniforth – Flickr profile: pstani)

Diet: A variety of small fruits, seeds and invertebrates ranging from woodlice to earthworms.

Identification: Small dense pellets that differ to similar-sized gravel or stones by their more spherical shape and textured appearance. In the autumn some may resemble a shrivelled blackberry. Vary in colour and structure depending on diet.

Where found: Fenceposts, footpaths, feeding stations and beneath favoured roosting or perching locations.

Size: Those from robin-size birds are small, the size of a sultana or currant or smaller. Pellets from song thrush and blackbird-size species may be larger, peanut-sized to the size of an apricot stone.

Notable features: Look for the appearance of loosely aligned small seeds (rather than the regularly aligned seeds of a shrivelled blackberry), beetle wing cases, legs of invertebrates such as spiders or flies and the greyer segments or plates of woodlice. These small pellets can easily be dismissed as poo, although lack the white, watery parts excreted by small songbirds. Instead, they have a distinct oval or sausage shape and are packed full of seeds, grit and the exoskeletons of invertebrates.

Two different robin pellets revealing the hard, indigestible exoskeleton remains of small invertebrates.

A robin pellet, enlarged and shown from different angles, revealing the hard body plates of invertebrates such as woodlice.

A pellet from a small insect-eating bird, from different angles, revealing small parts of invertebrate and an almost fully intact exoskeleton of a spider.

A blackbird-size pellet revealing a mass of fruit seeds, largely from blackberries (those with a pitted surface).

A wren-size pellet, less than 2.5mm long, recovered from the debris inside an old hollow tree. Shown lifesize (left) and enlarged (right). (Bob Cowley)

A brambling crop and its contents. This bird had been predated by a merlin. (Jack Ashton-Booth)

The contents of the brambling's crop: mostly comprising sea campion seeds and grit. (Jack Ashton-Booth)

10 OTHER SPECIES' PELLETS

There are a range of other species that produce pellets which are less likely to be found by people due to their secretive or aerial behaviour. For example, swifts, swallows and house martins are known produce pellets, although if regurgitated high up in the sky they are unlikely to be encountered (Tucker 1944; Duke 1977). Cuckoos also produce pellets, although again, thanks to their secretive nature and habitat, their pellets are less likely to be found. Cuckoos specialise in eating hairy caterpillars and have an unusually soft-walled gizzard for the hairs to embed into. When the time comes, the gizzard's membrane is shed as a pellet containing a mix of mucus and caterpillar hairs (Gill 1980). Although rare in Britain, across Europe hoopoes and rollers also produce pellets containing the remains of invertebrates, lizards and other animals. Two further groups of birds – shrikes and bee-eaters – produce pellets and while scarce in Britain and Ireland, do regularly occur and in some places may be encountered during the winter and summer, respectively.

Note: All images of pellets and pellet contents are shown life size unless a scale bar or a note in the photo caption indicates otherwise. Outdoor, 'in-the-field' images are not to scale.

Shrikes

Shrikes eat a variety of small animals, whose bones, hair or wing cases remain undigested and regurgitated as pellets. They are commonly photographed in the act of regurgitating pellets as these images – from shrikes in North America or Europe – reveal.

A loggerhead shrike regurgitating a pellet, Everglades, Florida, USA. (Steve Martz – Flickr profile: Swmartz)

A loggerhead shrike regurgitating a pellet, Palm Beach, Florida, USA. (John Sutton – Flickr profile: johnsutton580)

A red-backed shrike regurgitating a pellet while clutching an insect in its left claws. (Vince Garvey – Flickr profile: Baggie Bird 1)

A Turkestan shrike regurgitating a pellet. (Pete Rodgers – Flickr profile: Pete Rodgers)

Great grey shrike

In Britain and Ireland, the great grey shrike is a scarce yet regular winter visitor in small numbers (100 individuals or fewer). The red-backed shrike – now a rare breeding bird in Britain – still passes through in larger numbers during the autumn. Their pellets are most likely to be encountered near favoured perches. The profile below deals with the great grey shrike.

A great grey shrike pellet revealing mostly a small mammal's lower jawbone (with yellow incisor) and hair and a yellow feather from a blue tit. (Collected by Mike Dilger)

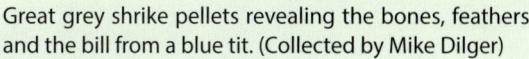

Great grey shrike pellets revealing the bones, feathers and the bill from a blue tit. (Collected by Mike Dilger)

A pellet from a northern grey shrike containing small mammal hair, Vinalhaven, Maine, USA. (Kirk Gentalen)

Pellet from a roosting Iberian grey shrike. It is packed full of small invertebrate remains and resembles a pellet-shaped muesli bar, Lanzarote, Canary Islands. (Bernhard Jacobi – Flickr profile: terraincognita96)

Diet: Invertebrates such as crickets and beetles, small birds, mammals and reptiles.

Identification: Grey to dark brown or black pellets depending on diet.

Where found: Below favoured perching trees and roosts. Nearby spiny tree branches or barbed-wire fencing may be host to whole or partial small birds, mammals or insects, left impaled.

Size: Small pellets, about the size of an apricot stone, 17–34mm long and 9–15mm thick (Nikolov 2004; Brown *et al.* 2021).

Notable features: The diet of a great grey shrike resembles that of a small falcon or little owl. The open, heath-type habitat and time of the year (late autumn to early spring), as well as presence of the bird(s), will help clinch identification.

European bee-eater

A male bee-eater regurgitating a pellet, Poiplie Special Protection Area, Slovakia. (Radovan Václav – Flickr profile: Radovan Václav)

Bee-eaters consume a range of flying insects and produce distinctive pellets full of the exoskeletons of their prey, particularly bees and wasps, often found below their nest burrows. Although a rare breeding species in Britain, bee-eaters are more commonly encountered on mainland Europe nesting in soft, sandy banks.

A selection of bee-eater pellets, revealing the variety of external body parts from invertebrates packed into them. Found beneath a nesting cliff, Grindul Lupilor, Romania.

Diet: A range of flying insects caught on the wing.

Identification: Dark, fibrous and delicate pellets packed full of the external hard remains of flying insects, particularly wings and heads.

Where found: Below in nest burrows and beneath nesting colonies and favourite perches. Visiting active bee-eater colonies may require a special licence (e.g. Schedule 1) or permit.

Size: 10–35mm long (Massa & Rizzo 2002).

Notable features: Regurgitate 4–6 times a day (Massa & Rizzo 2002). Most likely to contain wasps and bees, dragonflies and damselflies and smaller numbers of beetles, flies and other insects (such as butterflies and cicadas), which will vary in proportion throughout the seasons.

11 IDENTIFYING SMALL MAMMAL BONES

The bones – in particular the skulls – of small mammals such as voles, mice and shrews are some of the most common items encountered in pellets from a wide range of birds, including birds of prey, corvids and shrikes. The following extended chapter delves into detail on how to identify species of small and medium-sized mammals that are found in bird pellets across Britain and Ireland.

Note: All images of bones are shown life size unless a scale bar or a note in the photo caption indicates otherwise.

Wear and tear

When examining small mammal skulls extracted from bird pellets, many will be missing teeth while others may be from a mix of young and older individuals. The teeth of those from older animals may be more worn. For example, the lower incisor teeth of shrews may lack the bumps or cuspids that distinguish the various species. Therefore, using a range of clues, such as size, teeth pattern or number of teeth will help to clinch the skull's identification. Most skulls are likely to have damaged brain cases or craniums. This is where a bird of prey has crushed or pecked the skull to kill its prey or where it has been damaged during the digestion process. Some birds of prey may also break into the cranium to eat the nutritious brain first. Therefore, the pattern, shape and colour of the teeth – if present – are always important in identifying the skull.

Teeth

There are four different types of teeth found in mammal skulls: the canines, the incisors, the premolars and the molars. Different species of small to medium-sized mammals have these in different numbers, including in certain cases none at all for some types of teeth.

A dental formula is used to denote the type and number of teeth represented for different mammal species. It tells us how many types of teeth a mammal has on each of its upper and lower jaws. For instance, a vole's dental formula would be 1/1, 0/0, 0/0, 3/3 followed by a total number of 16 which represents all the teeth across all jaw bones. This formula tells us that there is one incisor on the upper and lower jaw, no canines or premolars, and three molars on the upper and lower jaws. Mice and rats also have the same dental formula, while dormice have 1/1, 0/0, 0/0, 4/4 (four molars on each jaw) and squirrels 1/1, 0/0, 0/0, 5/4 (five upper molars and four lower molars) (Harris & Yalden 2008).

On diagrams and in descriptions, teeth are often shortened to a single letter denoting their name and a superscripted number for teeth on the upper jaw and

subscripted for those on the lower jaw. A single incisor would thus be i¹ on the
upper jaw and i₁ on the lower jaw.

Voles, mice, dormice and squirrels are all rodents and have a distinctive pair
of incisors at the front of their upper and low jaws. Depending on the species
their colour may be ivory, yellow or orange. Rodents lack any canines and several
premolars. Instead of these teeth there is a gap, known as the diastema, followed
by three to five cheek teeth (premolars and molars). The molars are shortened to
m followed by a superscripted or subscripted number – so, the first molar on the
upper jawbone would be written as m¹ and on the lower jawbone as m₁.

For other small to medium-sized mammals whose remains may appear in pellets,
their teeth formula or descriptions can be found in their profiles later in this book.

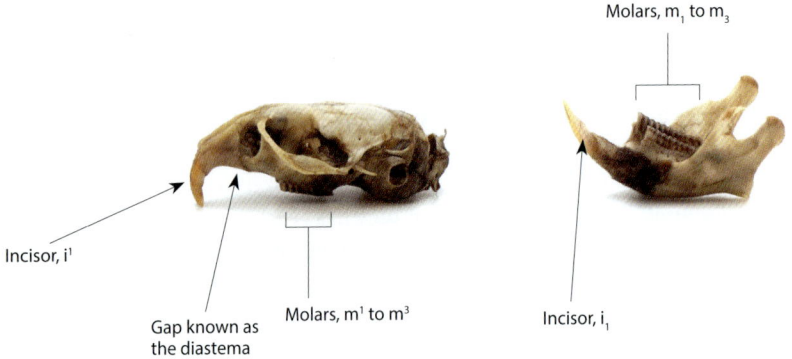

Molars, m_1 to m_3

Incisor, i^1

Gap known as
the diastema

Molars, m^1 to m^3

Incisor, i_1

The different teeth of an Orkney vole or common vole. Skull length is 30.1mm and
lower jaw length is 18.7mm. (Cat kill, collected by Stuart Williams)

Other bones

The skulls are the most easily identifiable part of voles and mice and are covered
in detail shortly. However, in among the skulls and hair there will be a variety of
other bones, mostly indistinguishable between species. Some of the common and
easily identified bones are shown below.

Pelvic bones

The pelvic bones of small mammals are very easy to spot. They
are a thin rod of bone with one end oval-shaped (see 'Sexing
small mammal bones' for more details).

[All images are life size unless otherwise indicated.]

Shoulder blade

The shoulder blade has a distinctive fan-like shape.

Back part of skull where vertebral column joins

Skulls often break apart into various fragments. The base of the skull, where it meets the neck bones or vertebrae, often comes away and this can cause confusion as to its identity. Here the fragment of skull shows part of the auditory bulla, a round bony structure that contains the middle and inner ear structures (left side of skull fragment).

Tibia and fibula

The shin bones, or tibia and fibula, together form a distinctive harp-like structure. The tibia is the thicker bone and the fibula forms the thinner, thread-like bone.

Ribs

The ribs are thin, gently curved bones found among the pellet debris.

Auditory bulla

The auditory bulla is a bony bulbous structure that contains the middle and inner ear structures. These easily detach from the skull of a mouse or vole and are often a mystery item. They are roundish in shape with a small hole (where the outer ear would connect).

Teeth

Teeth frequently fall out of jawbones, including the longer, curved yellow incisors.

A vole skull showing the hole pattern when all the teeth are missing (×20 magnification). (From a barn owl pellet, University of Bristol)

[All images are life size unless otherwise indicated.]

Sexing small mammal bones

The pelvic bones are shaped differently, depending on whether it is a vole, mouse or shrew and depending on whether they are from a male or female (Brown & Twigg 1969; Ronayne & Sleeman 2013). The photos below point to the key differences between male and female pelvic bones of each type of animal.

Field vole

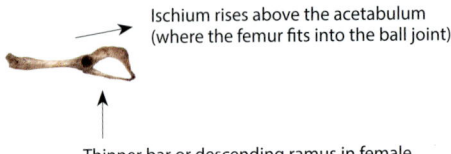

Ischium rises above the acetabulum (where the femur fits into the ball joint)

Thinner bar or descending ramus in female

The pelvic bone of a female field vole.

Broader ischium in male

The pelvic bone of a male field vole.

Wood mouse

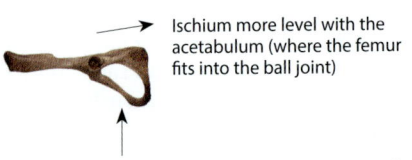

Ischium more level with the acetabulum (where the femur fits into the ball joint)

Thinner bar or descending ramus in female

The pelvic bone of a female wood mouse.

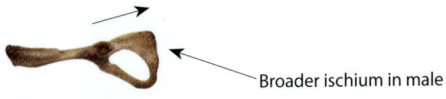

Broader ischium in male

The pelvic bone of a male wood mouse.

[All images are life size unless otherwise indicated.]

Common shrew

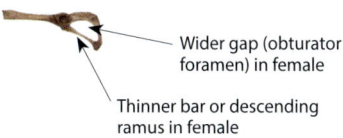

Wider gap (obturator foramen) in female

Thinner bar or descending ramus in female

The pelvic bone of a female common shrew.

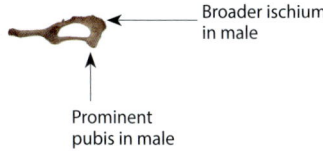

Broader ischium in male

Prominent pubis in male

The pelvic bone of a male common shrew.

Pygmy shrew

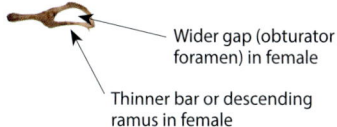

Wider gap (obturator foramen) in female

Thinner bar or descending ramus in female

The pelvic bone of a female pygmy shrew.

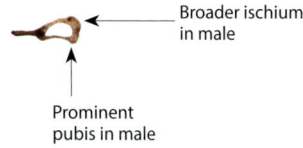

Broader ischium in male

Prominent pubis in male

The pelvic bone of a male pygmy shrew.

[All images are life size unless otherwise indicated.]

Voles

When you open up an owl pellet, voles – in particular field voles – are likely to be the most common small mammal prey found. Where they occur, which is across most of mainland Britain, field voles are the staple prey for barn, tawny, long-eared and short-eared owls. They are also eaten by other birds of prey, shrikes, herons and corvids.

Voles are small rodents and there are four species that may be encountered: 1) the field or short-tailed vole, 2) the bank vole (and its subspecies), 3) the water vole, and 4) the Orkney, Guernsey or common vole. The only one occurring on the island of Ireland is the introduced bank vole.

When you look closely at small mammal skulls from owl pellets, vole skulls with teeth are easily identified. The teeth form a zigzag pattern with flattened triangular surfaces.

The length of vole skulls can be variable depending on whether they were male, female or younger animals. The male voles are generally larger than females, and field voles are larger than bank voles. Although owls eat voles whole, the cranium part of the skull is usually damaged from where it has been pecked at by the owls to kill or through the mechanical and chemical damage caused while in the stomach.

Field and bank vole skulls are very similar. The teeth help identify whether the species in question is a bank vole or a field vole. If a tooth is plucked out from the skull or jaw and it is a continuous tooth leading into the root, it is most likely to be from a field vole. However, if the lower portion has two roots (like two small prongs) then it will be from a bank vole. Young bank voles, less than three months of age, will have less-developed roots and so theirs may resemble field vole teeth. As mentioned earlier, all voles are missing their canines and premolars: each upper and lower jaw has one incisor, then a gap (the diastema) followed by three cheek teeth (molars) (McDonald & Barrett 1993).

Field vole

Field vole skulls are the most common skull found in owl pellets on mainland Britain. Elsewhere, depending on presence and distribution, bank vole and Orkney vole may dominate. The field vole is absent from Ireland, some Scottish islands including Lewis, Barra, some islands in the Inner Hebrides, Orkney, Shetland, Ireland, Isle of Man, Lundy, Isles of Scilly and all of the Channel Islands (Harris & Yalden 2008). When dissecting pellets, they are generally the larger of the common small mammal skulls encountered (excluding rat and water vole). Some pellets, especially those from barn owls, may contain three, four or more skulls.

A selection of field vole skulls. (From barn owl pellets, University of Bristol)

The undersides of three field vole skulls and the variation in their condition. (From barn owl pellets, University of Bristol)

The lower-left jawbones of two field vole skulls showing their outside and inside edge. (From barn owl pellets, University of Bristol)

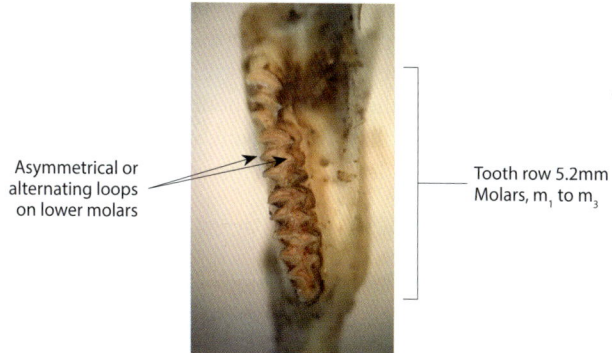

Asymmetrical or alternating loops on lower molars

Tooth row 5.2mm
Molars, m_1 to m_3

A field vole lower-left jawbone showing the teeth pattern and the asymmetrical pattern of loops on the teeth where each loop on one side of the tooth alternates with each one on the other side. The bottom of the photo is the front (anterior) or nose-end of the skull. (From a barn owl pellet, University of Bristol)

[All images are life size unless otherwise indicated.]

3mm

A field vole tooth showing the single continuous tooth.
(From a barn owl pellet, University of Bristol)

Additional loop or nodule on the upper middle molar, m²

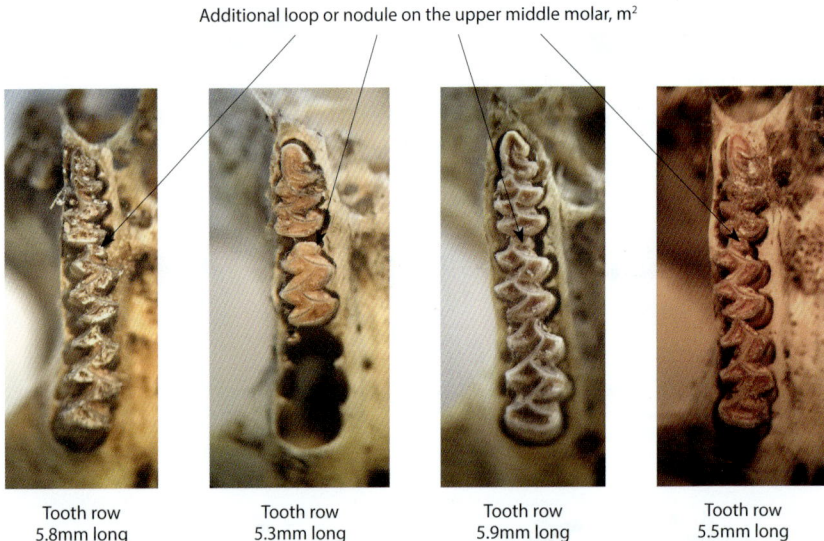

| Tooth row 5.8mm long | Tooth row 5.3mm long | Tooth row 5.9mm long | Tooth row 5.5mm long |

Field vole upper-left teeth rows from four different individuals. The bottom of each photo is the front (anterior) or nose-end of the skull. (From barn owl pellets, University of Bristol)

Quick identification: Large small-mammal skull. Very common, if not the most common prey item in barn owl pellets and those from other birds of prey. Field voles have a set of fused, continuous molars that have a single root. They can easily be viewed in full if plucked from the skull or jawbones using tweezers. If plucked molar teeth have two roots they will be from a bank vole (note: young bank voles (up to 3 months old) have not yet developed the two roots).

Skull length range: 23–29mm (Macdonald & Barrett 1993).

Clincher: The middle molar (m^2) has an extra loop or nodule that is missing from the bank vole and Orkney vole. Molar teeth have sharp angles compared to those of bank voles. Teeth row is less than 7mm, shorter than water voles which are more than 9mm (Harris & Yalden 2008). The loops on each side of the lower molar teeth are asymmetrical and alternating (Yalden 2009).

[All images are life size unless otherwise indicated.]

Bank vole

Depending on the bird of prey species, habitat and location, remains of the bank vole may be commoner in pellets than those of the field vole. It is found across mainland Britain and on many of the islands (probably introduced), where populations exhibit subtle differences in morphology, such as hair colour, more complex molar shape in m³ and larger skull and hindfoot lengths. While these differences, compared to the mainland population, mean they are generally referred to as island subspecies, such as the Skomer vole, the Mull vole and the Jersey vole, they may in fact just be populations at the extreme end of their size variation (Harris & Yalden 2008). The bank vole is the only vole in Ireland, found in the Republic of Ireland only where it was probably accidentally introduced. It is absent from Northern Ireland, Orkney, Shetland, Isle of Man, most of the Channel Islands apart from Jersey, and many islands in the Outer and Inner Hebrides (Harris & Yalden 2008).

Bank voles use more woodland and enclosed habitats compared to the field vole, and may therefore be less available to birds of prey hunting over rough grassland. They are probably hunted along hedgerows, in woodlands and in vegetation buffering woodland and copses.

A selection of bank vole skulls. (From barn owl pellets)

Two Jersey bank vole skulls. (From barn owl pellets, collected by the Jersey Barn Owl Conservation Trust and Ian Buxton)

The underside of two bank vole skulls. (From barn owl pellets, University of Bristol)

[All images are life size unless otherwise indicated.]

The underside of a Jersey bank vole skull. (From a barn owl pellet, collected by the Jersey Barn Owl Conservation Trust and Ian Buxton)

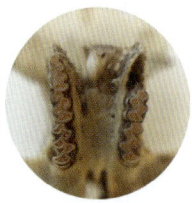

The underside of a bank vole skull, focusing on the teeth pattern (×20 magnification). The bottom of the photo is the front (anterior) or nose-end of the skull. (From barn owl pellets, University of Bristol)

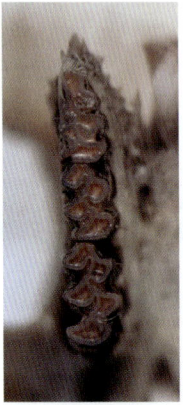

Tooth row
5.4mm long

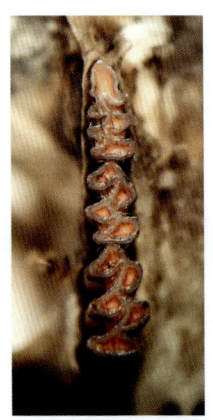

Tooth row
5.1mm long

Bank vole upper-left teeth rows from two different individuals. The bottom of each photo is the front (anterior) or nose-end of the skull. (From barn owl pellets, University of Bristol)

The lower-left jawbones of two Jersey bank vole skulls. (From barn owl pellets, collected by the Jersey Barn Owl Conservation Trust and Ian Buxton)

[All images are life size unless otherwise indicated.]

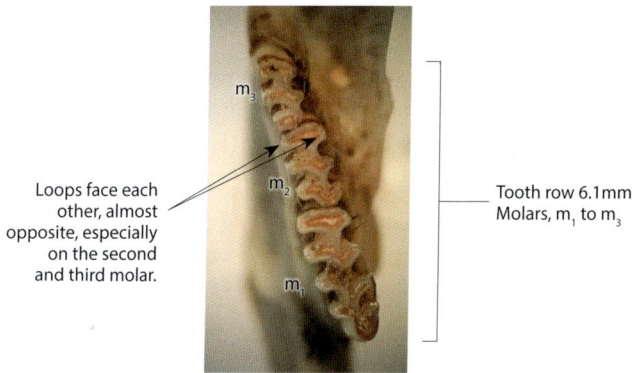

Loops face each other, almost opposite, especially on the second and third molar.

m_3

m_2

m_1

Tooth row 6.1mm
Molars, m_1 to m_3

A Jersey bank vole lower-left jawbone showing the teeth pattern and the generally symmetrical pattern of loops on the teeth where each loop on one side of the tooth is opposite those on the other side. The bottom of the photo is the front (anterior) or nose-end of the jawbone. (From a barn owl pellet, collected by the Jersey Barn Owl Conservation Trust and Ian Buxton)

2.5mm

A bank vole tooth showing the two roots. (From a barn owl pellet, University of Bristol)

Quick identification: Pluck a tooth and look for two roots rather than one continuous tooth.

Beware, though, that younger bank vole teeth (of individuals less than three months old) will lack the long roots, although may have hints of them. These skulls may be smaller and, if the teeth are present, can be identified by their teeth pattern using a hand lens or microscope (or zooming in on a tablet or visualiser). These are most likely to have come from pellets found during the bank vole's breeding season (April to September/October). Be aware if collecting pellets in the winter as they are likely to be from across the seasons and include the skulls of young bank voles. This will help avoid misidentifying the remains as field vole (MacDonald & Barrett 1993; Harris & Yalden 2008).

Skull length range: 20.3–24.9mm (Macdonald & Barrett 1993, Harris & Yalden 2008).

Clincher: The teeth pattern will help confirm the species as bank vole. The middle upper tooth (m^2) will lack the extra kink or nodule found in the field vole. Compared to field voles, their molar teeth have a smoother surface and the cranium is less angular (Harris & Yalden 2008). The loops on each side of the lower molar teeth are largely symmetrical and opposite each other (Yalden 2009).

Orkney, Guernsey or common vole

In Britain, the common vole – which occurs across Europe – is found on eight islands in Orkney and on Guernsey in the Channel Islands, where it is known as the Orkney vole and Guernsey vole respectively (Harris & Yalden 2008).

Two Orkney vole skulls. (Cat kills, collected by Stuart Williams)

Tooth row
6.1mm long

Tooth row
6.3mm long

Tooth row
6.1mm long

Orkney vole upper-left teeth rows from three individuals. The bottom of each photo is the front (anterior) or nose-end of each skull. (Cat kills, collected by Stuart Williams)

Quick identification: This is the only vole species found on Orkney and Guernsey.

Maximum skull length varies between geographic location. Up to 30mm on Orkney mainland, Rousay and South Ronaldsey; 29mm on Sanday and Westray; 28mm on Guernsey and 24mm in lowland Western Europe (Harris & Yalden 2008).

Clincher: The middle upper tooth (m^2) lacks the extra kink or nodule found in the field vole.

Comparing vole teeth patterns

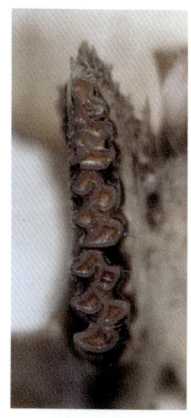

Tooth row
5.5mm long

Tooth row
5.4mm long

Tooth row
6.1mm long

These photos provide a comparison of the pattern of teeth in each vole species (upper-left teeth row). Field vole (left), bank vole (middle) and Orkney vole (right). The field vole has an extra kink or nodule on the middle tooth (m^2). The bottom of each photo is the front (anterior) or nose-end of each skull.

Water vole

In some locations, particularly where there are waterways (such as ditches, rhynes or wetland habitats), water voles may be eaten by predators such as barn owls, grey herons and marsh harriers.

The water vole is much larger than the bank or field vole, with broader, larger zig-zag patterned teeth. At first glance, their skull could be confused with that of a brown rat. However, the teeth and the pattern they leave behind (if missing) should confirm the identification.

The water vole has been fast declining across Britain and is absent from Ireland. See the National Water Vole Database & Mapping Project for their latest distribution.

An adult water vole skull. (From a barn owl pellet, University of Bristol)

[All images are life size unless otherwise indicated.]

The underside of a water vole skull. (From a barn owl pellet, University of Bristol)

The front view of an adult water vole skull, revealing the orange teeth. (From a barn owl pellet, University of Bristol)

An adult water vole lower right jawbone. (From a barn owl pellet, University of Bristol)

A smaller water vole skull from a young animal. It is larger than a field vole skull with darker orange teeth, although the teeth row in this specimen is only just over 7mm in length (at 7.33mm). Identification can be confirmed by the lack of a loop or nodule on the middle molar (m^2). (Collected by Dr Jim Vafidis)

Quick identification: Large rat-size skull (average 41mm long from front of skull to back) with zig-zag patterned molar teeth and bright orange incisors (Harris & Yalden 2008).

Skull length range: For females 37.2–42.5mm and for males 39–44.4mm (Macdonald & Barrett 1993). Shorter skull length in other parts of Europe (Harris & Yalden 2008).

Clincher: On adults, teeth row is larger than 9mm. Teeth grow as a single continuous unit like field and common voles, lacking any fixed roots (Harris & Yalden 2008). Middle molar (m^2) is missing the loop or nodule seen in the field vole.

[All images are life size unless otherwise indicated.]

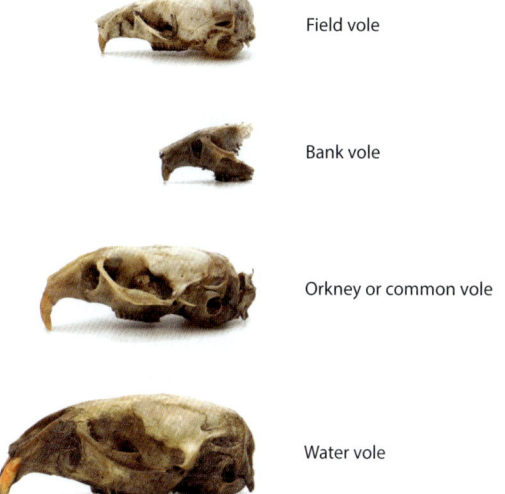

Field vole

Bank vole

Orkney or common vole

Water vole

A size comparison of the four species of vole that may be encountered in pellets. The bank vole skull has the brain case (cranium) damaged and missing.

Mice and rats

Mice and rats are common prey for many bird predators and are found in the pellets of owls, eagles, buzzards, harriers, shrikes, corvids and gulls. There are six species that may be encountered: 1) the wood mouse, 2) the house mouse, 3) the yellow-necked mouse, 4) the harvest mouse, 5) the brown rat, and the 6) black rat. The wood mouse is a very common prey item and is the one most often encountered in pellets.

At first glance, mouse and rat skulls look very similar in shape to those of voles and can be identified by the absence of zig-zag patterned molar teeth. Instead, the skulls of mice and rats have teeth with cusps that form ridges and troughs, a little like our own teeth. The rows of cusps look like a pattern of tightly packed spheres or points. As in voles, their canines and premolars are missing. Therefore, on each side of both the upper and jaw there is one incisor, then a gap (the diastema) followed by three cheek teeth (molars) (McDonald & Barrett 1993).

Whether the teeth are present or absent, the arrangement of holes that the teeth fit into are the most reliable ways of identifying a mouse to species. In particular, the first molar (m^1) closest to the nose-end of the skull will have a different number of root holes depending on the species. For example, the wood mouse has four holes while the harvest mouse has five. If the teeth are present, they can be plucked using tweezers to reveal the root holes for that tooth. Where the teeth are already missing, the position and pattern of the holes should still help with identification. Rat skulls are much larger, up to twice as long depending on the age of the animal, and have cusped teeth unlike the water vole. Smaller rat skulls can be confirmed by their teeth root pattern.

The pattern of cusps on the teeth, especially on m^1, can also confirm identification. For instance, the cusps of the wood mouse are symmetrical, while those of the house mouse are asymmetrical.

The upper incisors of wood mice and house mice differ in their shape where they have a cutting edge or notch. Those of wood mice have a simple, angled cutting edge, while those of house mice have an extra cutting edge or notch.

Wood mouse

Alongside the field vole, the wood mouse is one of the most common prey items encountered in owl pellets. It is found across most of Britain and Ireland, including islands, although avoids open, mountainous areas (Harris & Yalden 2008).

Thickness of incisor, front to back, measures 1.1–1.3mm

An enlarged wood mouse skull. (From a barn owl pellet, University of Bristol)

These photos show the range in condition of wood mouse skulls when extracted from pellets. (From barn owl pellets, University of Bristol)

The underside of four wood mouse skulls. (From barn owl pellets, University of Bristol)

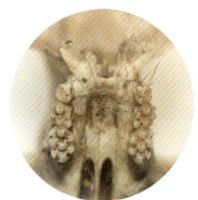

A close-up of the underside of a wood mouse skull, showing both teeth rows and the cusped teeth (×20 magnification). The bottom of the photo is the front (anterior) or nose-end of the skull. (From a barn owl pellet, University of Bristol)

[All images are life size unless otherwise indicated.]

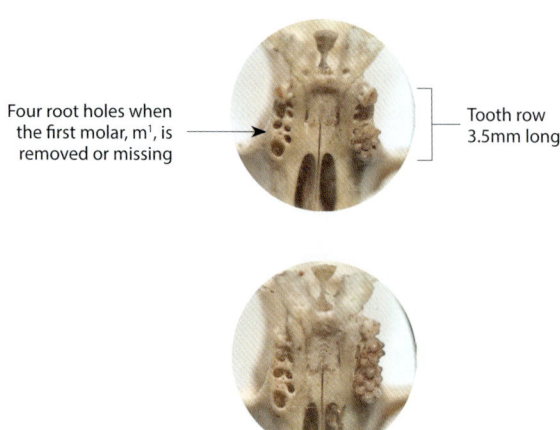

Four root holes when the first molar, m¹, is removed or missing

Tooth row 3.5mm long

The underside of wood mouse skulls, showing the distinctive root hole patterns on the first molar (m¹). The bottom of each photo is the front (anterior) or nose-end of the skull. The right-hand photo above shows a close-up of the upper-left jaw without any teeth. (From barn owl pellets, University of Bristol)

The lower-left jawbone of a wood mouse, showing the outside and inside edge. (From a barn owl pellet, University of Bristol)

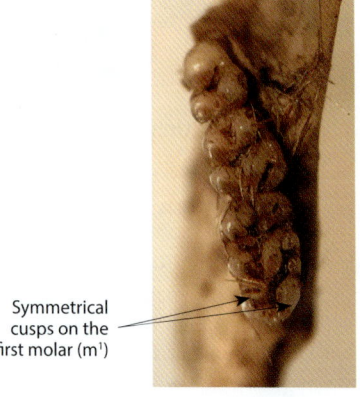
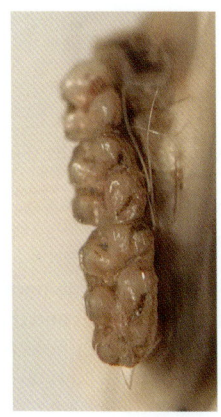

Symmetrical cusps on the first molar (m¹)

Tooth row 3.7mm long   Tooth row 4mm long   Tooth row 3.8mm long

Lower jawbones of three wood mice, showing the symmetrical cusps on the first molar (m¹). The bottom of each photo is the front (anterior) or nose-end of each jawbone. (From barn owl pellets, University of Bristol)

[All images are life size unless otherwise indicated.]

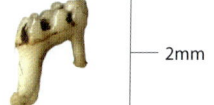

— 2mm   A wood mouse first upper molar (m¹) tooth revealing the cusps and multiple roots. (From a barn owl pellet, University of Bristol)

Quick identification: Alongside field vole, wood mice skulls are one of the commonest skulls found in owl pellets. They have cusped teeth rather than the zig-zag patterned teeth of voles.

Skull length range: 22–26mm (Macdonald & Barrett 1993). Upper cheek teeth row 3.6–4.1mm (Harris & Yalden 2008).

Clincher: Four (rarely five) root holes in the upper jaw where the first tooth (m^1) sits, closest to the incisor tooth and nose-end of the skull. This tooth also has symmetrical cusps. Length between m^1 to m^3 ranging between 3.7–4.6mm, so longer than house mouse or yellow-necked mouse (Harris & Yalden 2008). Although very similar to scarcer species the yellow-necked mouse, the incisors are less thick and from the front (anterior) to the back (posterior) measure 1.1–1.3mm. The lower jawbone is also shorter (12.3–16.4mm long) (Harris & Yalden 2008). Wood mouse skulls can be distinguished from those of the house mouse by their lack of a notch on the upper incisor teeth.

[All images are life size unless otherwise indicated.]

House mouse

The house mouse is a common and widespread non-native rodent, found across Britain and Ireland. At first glance its skull is very similar to the wood mouse and the yellow-necked mouse. However, subtle differences in the teeth should help to distinguish it.

A skull of a house mouse revealing the notched front teeth or incisors. (Bristol Museum & Art Gallery)

Another skull of a house mouse, again highlighting the features of the teeth, Western Brittany, France. (Collected by Bernard Cadiou)

The underside of a house mouse skull. (Bristol Museum & Art Gallery)

A house mouse's front teeth (enlarged), revealing the characteristic notched incisor. (Bristol Museum & Art Gallery)

A close-up of a house mouse's upper-left teeth row. The bottom of the photo is the front (anterior) or nose-end of the skull. (Bristol Museum & Art Gallery)

A close-up of a house mouse's upper-left teeth row. The bottom of the photo is the front (anterior) or nose-end of the skull, Western Brittany, France. (Collected by Bernard Cadiou)

[All images are life size unless otherwise indicated.]

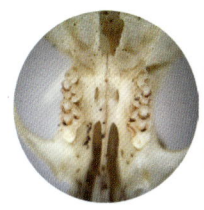

A close up of a house mouse's upper-left and upper-right rows of teeth (×20 magnification). The bottom of the photo is the front (anterior) or nose-end of the skull, Western Brittany, France. (Collected by Bernard Cadiou)

The lower-left jawbone of a house mouse. (Bristol Museum & Art Gallery)

The jawbone of a house mouse tilted to show the cusped teeth rows. (Bristol Museum & Art Gallery)

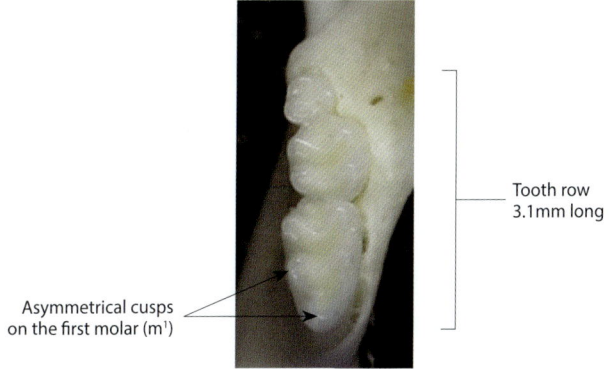

Tooth row 3.1mm long

Asymmetrical cusps on the first molar (m^1)

A house mouse lower-left jawbone with asymmetrical cusps on tooth (m^1). The bottom of the photo is the front (anterior) or nose-end of each jawbone. (Bristol Museum & Art Gallery)

Quick identification: House mice have the cusped teeth typical of other mice. However, they also have a distinctive notched incisor that is unlike other mice found in Britain and Ireland.

Skull length range: 20–25mm (Macdonald & Barrett 1993). Cheek teeth row 2.9–3.4mm (Harris & Yalden 2008).

Clincher: Three root holes in the upper jawbone where the first molar (m^1) sits and asymmetrical cusps on the lower first molar (m_1) (Harris & Yalden 2008). Length of m^1 to m^3 ranges between 2.9 and 3.4mm. The third molar (m^3) is distinctly smaller than the other two (more so than in wood mouse, yellow-necked mouse or harvest mouse).

[All images are life size unless otherwise indicated.]

Yellow-necked mouse

The yellow-necked mouse is generally less common than wood mice – although it may be locally common – and has a more restricted range, found across parts central and eastern Wales, counties bordering Wales and the south and east of England (Harris & Yalden 2008; Mathews *et al.* 2018). This species is absent from Ireland.

It is generally a larger mouse than the wood mouse, although there may be some overlap with younger individuals. The skull is incredibly similar, although in adults will be slightly larger than that of the wood mouse.

Thickness of incisor, front to back, measures 1.45–1.65mm

An enlarged yellow-necked mouse skull. (Bristol Museum & Art Gallery)

Another skull of a yellow-necked mouse. (Collected by Dr Rosie Plummer)

The underside of a yellow-necked mouse skull. (Bristol Museum & Art Gallery)

[All images are life size unless otherwise indicated.]

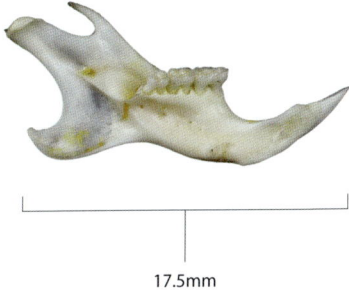

17.5mm

A yellow-necked mouse lower-left jaw (inside edge). This jaw measures 17.5mm, safely within the range of this species (compared to wood mouse). (Bristol Museum & Art Gallery)

A pair of yellow-necked mouse jawbones. (Collected by Dr Rosie Plummer)

Quick identification: Yellow-necked mice are scarce both in terms of their distribution and their appearance in owl pellets. If the lower jaw is 16.5mm or longer, then it will be yellow-necked mouse (16.0–18.8mm). That of the wood mouse measures 12.3–16.4mm.

Skull length range: 25–29mm (Macdonald & Barrett 1993).

Clincher: The upper incisors are thicker than in wood mice. Measured from the front of the tooth to the back, those of yellow-necked mice will be between 1.45 and 1.65mm. There may be overlap between the species: if they measure between 1.35 and 1.40mm they could be either species – a young yellow-necked mouse or an adult wood mouse with worn teeth (Harris & Yalden 2008). They have symmetrical cusps on their lower first molars (m_1).

[All images are life size unless otherwise indicated.]

Harvest mouse

The harvest mouse is encountered much less frequently in pellets and is generally an uncommon find, although depending on the location, may be locally common in some bird diets, particularly those of barn owls and kestrels. The species has a limited distribution in Britain and is absent from Ireland. It is mostly found in southern, central and eastern England, with the range extending into parts of Wales and the north-east of England (Harris & Yalden 2008; Mathews 2018).

Two harvest mouse skulls. (From barn owl pellets, University of Bristol)

The underside of a harvest mouse skull with the upper-left molars missing. (From a barn owl pellet, University of Bristol)

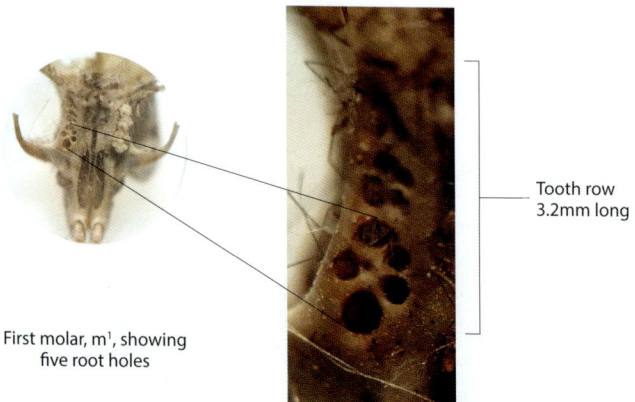

Tooth row 3.2mm long

First molar, m¹, showing five root holes

A closer view of the underside of a harvest mouse skull. It reveals the root hole pattern, in particular five holes where the first molar would be. This is shown more clearly in the magnified photo showing only the tooth row. The bottom of the photo is the front (anterior) or nose-end of the skull. (From a barn owl pellet, University of Bristol)

Three harvest mouse lower jawbones. (From barn owl pellets, University of Bristol)

[All images are life size unless otherwise indicated.]

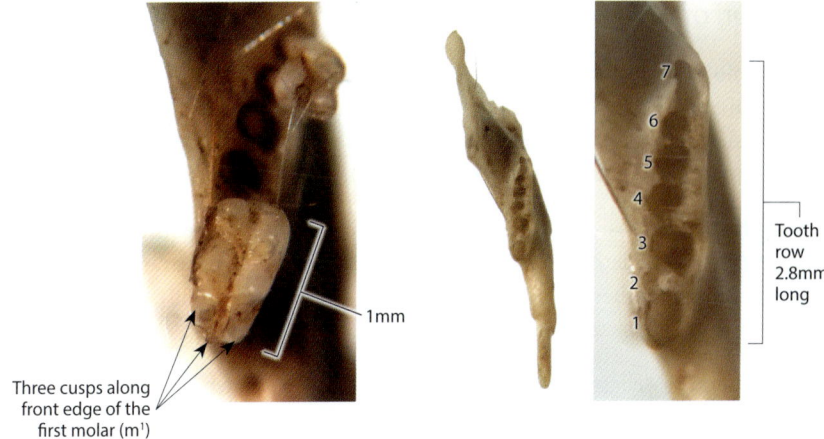

Three cusps along front edge of the first molar (m¹)

1mm

7
6
5
4
3
2
1

Tooth row 2.8mm long

Harvest mouse lower-left teeth row with the middle molar, m_2, missing. The bottom of the photo is the front (anterior) or nose-end of the jawbone. (From a barn owl pellet, University of Bristol)

Harvest mouse lower-left teeth row with all the teeth missing, revealing the seven root holes, three for m_1, two for m_2 and two for m_3. The bottom of the photo is the front (anterior) or nose-end of the jawbone. (From a barn owl pellet, University of Bristol)

Quick identification: Small, delicate and compact skull with a short muzzle and cusped teeth. No notched incisor (Harris & Yalden 2008). Lower jawbones short and compact with simple, angled joint. The brain case is often broken or absent.

Skull length range: 16–18mm (Macdonald & Barrett 1993). Cheek teeth row 2.6–2.8mm (Harris & Yalden 2008).

Clincher: Five root holes in each upper jaw where the first molar (m¹) sits and symmetrical cusps on the same tooth. The third molar (m³) has three root holes, the middle of which is very thin (Harris & Yalden 2008).

Common or brown rat

The common or brown rat is a ubiquitous non-native rodent found across Britain and Ireland, although less so in exposed mountainous areas and is absent from some smaller islands (and where successful rat eradication programmes have been carried out to help seabird populations recover).

Although adults are large, younger rats are closer in size to large mice and appear in the pellets of predators from grey herons to barn owls. Larger rats may still be hunted and dismembered or scavenged as carrion (from roadkill). Poisoned rats may also be easier to catch or find once dead.

When dissecting pellets, the larger teeth, skulls and lower jawbones of rats quickly become obvious among vole, mice and shrew skulls.

A skull of a brown rat. (Bristol Museum & Art Gallery)

Another skull of a brown rat. (Bristol Museum & Art Gallery)

The underside of a brown rat skull. (Bristol Museum & Art Gallery)

The underside of a further two brown rat skulls. (Bristol Museum & Art Gallery)

[All images are life size unless otherwise indicated.]

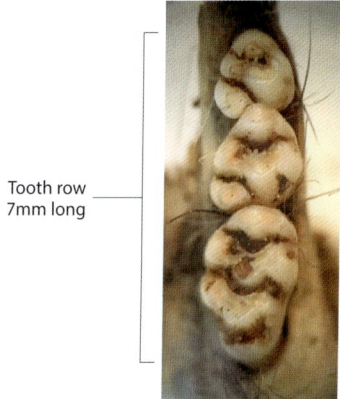

Tooth row
7mm long

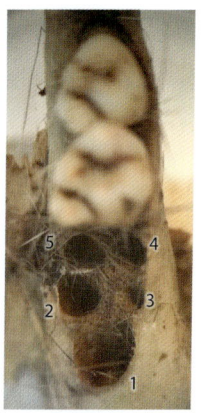

Five root holes
when the first molar
(m¹) is missing

The teeth on the upper-left jaw of a brown rat (left photo). The same upper-left jawbone of a brown rat revealing the root pattern when the first molar (m¹) is removed (right photo). The bottom of each photo is the front (anterior) or nose-end of the skull. (From a barn owl pellet, collected by the Jersey Barn Owl Conservation Trust and Ian Buxton)

A lower-left jawbone of a young brown rat showing the three molar teeth. The front (anterior) or nose-end is on the left, where the long incisor can be seen. (Bristol Museum & Art Gallery)

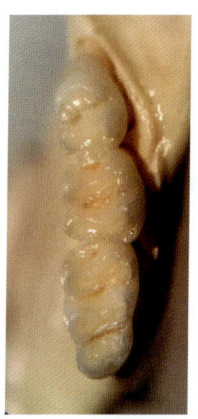

Tooth row 7mm long

The lower-left jawbone of a different brown rat showing the three molar teeth. The bottom of the photo is the front (anterior) or nose-end of the jawbone. (Bristol Museum & Art Gallery)

[All images are life size unless otherwise indicated.]

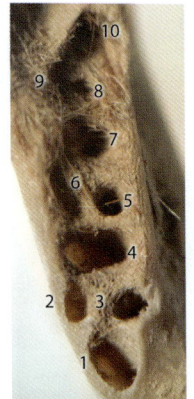

Ten root holes in total when all teeth are removed or missing

Tooth row 7mm long

Tooth row 8mm long

A close-up of the root holes in two brown rat lower-left jawbones. The bottom of each photo is the front (anterior) or nose-end of each jawbone. (Bristol Museum & Art Gallery)

Fibula

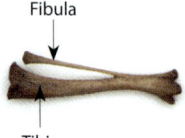

Tibia

The leg bones of a young brown rat: the tibia (shin bone) and fibula. (From a barn owl pellet, collected by the Jersey Barn Owl Conservation Trust and Ian Buxton)

The pelvis bone from a young brown rat. (From a barn owl pellet, collected by the Jersey Barn Owl Conservation Trust and Ian Buxton)

Quick identification: Flatter, broad, ridged teeth with less distinct cusps and larger in size – including from young animals – compared to mice, voles and shrews.

Clincher: Five root holes in each upper jaw where the first molar (m^1) sits. Three cheek teeth in each jawbone (rather than five as found in squirrels). Length of skull in large adults up to 54mm, range 43–54mm (Macdonald & Barrett 1993). The body weight of rats can be established by measuring the jaw length. See Yalden (2008) for graph and formula to establish weight.

[All images are life size unless otherwise indicated.]

Black or ship rat

The black rat, also known as the ship rat, was once a common and widespread non-native species across Britain and Ireland, brought here via trade ships from around the world. It often ended up on islands through shipwrecks and the moving of livestock. It has been largely replaced by the brown rat and eradicated on many offshore islands to protect seabird populations.

While exact numbers of black rats across mainland Britain are uncertain (Mathews *et al.* 2018), a recent study reveals black rats still occurring at ports or docks (Harris 2022). They still exist in the Republic of Ireland, for example on Lambay Island, on the Channel Islands of Alderney, Herm and Sark and Inchcolm Island, close to Edinburgh, in the Firth of Forth, Scotland.

A black rat skull. (Bristol Museum & Art Gallery)

The underside of a black rat skull. (Bristol Museum & Art Gallery)

Quick identification: Flatter, broad, ridged teeth with more distinct cusps than brown rat and larger in size compared to mice, voles and shrews.

Skull length range: 38–43mm (Macdonald & Barrett 1993).

Clincher: Five root holes in the first molar (m^1) and three cheek teeth (rather than five found in squirrels). Generally smaller, lighter and more angular skull than brown rat, with a rounded brain case (Harris & Yalden 2008).

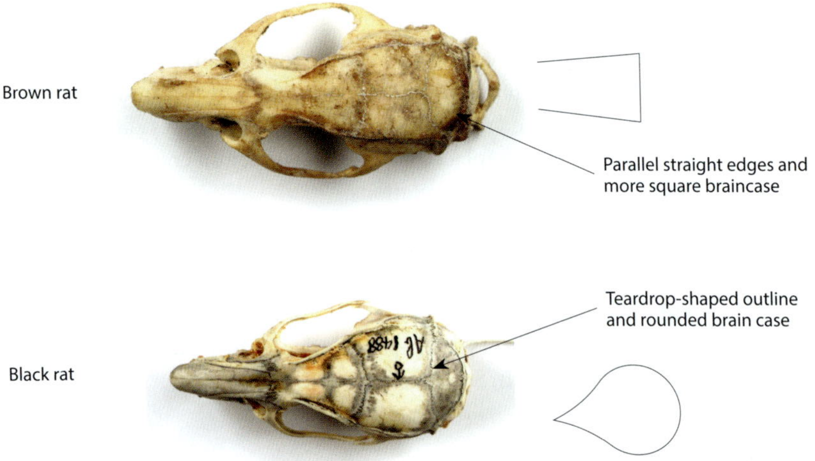

Brown rat

Parallel straight edges and more square braincase

Teardrop-shaped outline and rounded brain case

Black rat

Comparison of the shape of the brain case in the brown rat (parallel-sided) and black rat (rounded) from above.

Squirrels

Red and grey squirrels are common prey for larger bird predators such as goshawks, and occasionally peregrines and tawny owls, while buzzards, red kites and corvids such as ravens and crows will scavenge those that have been killed by cars. While squirrel skulls may appear in pellets, their hair and claws are more likely to be found.

Squirrel skulls look similar to other rodents and are large like those of rats and water voles. However, instead of three cheek teeth on each jawbone, they have five on their upper jawbones and four on their lower jawbones. As with other rodents, their canines are missing. In their upper jaw, the first tooth you see (from the nose-end) is an incisor, then a gap (the diastema), followed by a tiny premolar (p^3), a large premolar (p^4), followed by three molars (m^1 to m^3). On the lower jaw they have one incisor, the diastema, one premolar (p_4) and three molars (m_1 to m_3) (McDonald & Barrett 1993; Harris & Yalden 2008).

Red squirrel

The red squirrel has a limited range in Britain and Ireland, and where it is found it is an important food source for birds of prey such as goshawks. In Britain the red squirrel has mostly a northerly distribution, often as isolated populations in Wales and northern England, although can be encountered in southern England on Jersey (introduced), Isle of Wight and on Brownsea and Furzey islands in Dorset. It is found on Anglesey and absent from the Isle of Man. In Scotland it is more widespread, although absent from most islands (Harris & Yalden 2008; Mathews 2018).

A red squirrel skull. (Bristol Museum & Art Gallery)

The underside of a red squirrel skull. (Bristol Museum & Art Gallery)

[All images are life size unless otherwise indicated.]

A pair of red squirrel jawbones from above and from the left side. (Bristol Museum & Art Gallery)

Quick identification: The distribution of the red squirrel should help with identification. It has a large, rat-size rodent skull with five upper cheek teeth (the first, p^3, very small). The skull is wider and less elongated than that of a rat.

Clincher: The skull and teeth are larger and broader that those of rats, with low ridges around their edges (rather than running across the teeth). Compared with a grey squirrel, red squirrel skulls are smaller, more compact and have a shorter muzzle (nasal bones). Skulls are less than 50mm in length, with a range of 44–48mm (Macdonald & Barrett 1993, Harris & Yalden 2008).

Grey squirrel

The grey squirrel is a common non-native rodent found across England and Wales, and some parts of Scotland and Ireland (Harris & Yalden 2008; Mathews 2018). It is absent from most islands. Where it replaces the red squirrel, it has also become important hunted prey for birds such as goshawks and as scavenged prey for buzzards, red kites and corvids.

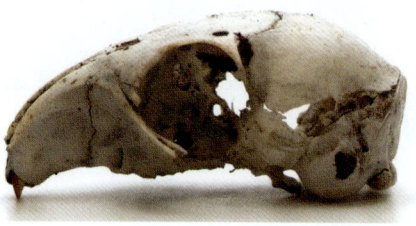

Two grey squirrel skulls.

[All images are life size unless otherwise indicated.]

This skull has the full complement of teeth in the teeth rows apart from the premolars (p³).

The underside of three grey squirrel skulls and a closer view of the upper left teeth row (p³ missing and m³ just out of view; bottom of photo is the front (anterior) or nose-end).

Quick identification: The distribution of the grey squirrel should help with identification. It has a large rodent skull with five cheek teeth (the first, p³, very small).

Skull length: Up to 56mm (Macdonald & Barrett 1993).

Clincher: The skull and teeth are larger and broader that those of rats with low ridges around their edges (rather than running across the teeth). Compared with red squirrel, those of grey squirrels are larger and more elongated with a longer muzzle (nasal bones) (Harris & Yalden 2008).

[All images are life size unless otherwise indicated.]

Dormice

Hazel dormice remains are rarely or only occasionally found in pellets in Britain. In other parts of their range, dormice species, including the hazel dormouse and edible dormouse, may be eaten by bird predators such as tawny owls, barn owls, long-eared owls and eagle owls (Anděra & Horacek 1986; Obuch 2001).

Dormice have small delicate skulls and distinctive flattened teeth with subtle, low ridges across them. They are missing any canines and premolars 1 to 3 (p¹ to p³). However, unlike voles and mice of a similar size, they have an extra tooth – a premolar – giving them four cheek teeth. Therefore, each side of both the upper and jaw contains one incisor, the premolar (p⁴) and three molars (m¹ to m³) (McDonald & Barrett 1993).

Hazel Dormouse

Hazel or common dormice are elusive and vulnerable to extinction in Britain. They are absent from Ireland. Their rarity makes them a less common feature in pellets of birds such as tawny owls, in which they do appear from time to time. They are widespread in patches across southern England (especially the west and southwest) and into Wales, with more fragmented populations in the Midlands and eastern England, and isolated populations further north (Harris & Yalden *et al.* 2008; Mathews 2018).

The underside of the remains of a dormouse skull showing the palate and some teeth. (From a tawny owl pellet)

A dormouse lower-left jawbone revealing the distinctive hole in the hinge base and the broad flattened teeth with low ridges. This jawbone has the full complement of teeth in the teeth row apart from the premolar (p_4). The first molar (m_1) is longer than the other two molars, m_2 and m_3. (From a tawny owl pellet)

A dormouse lower left jawbone, again showing the distinctive hole in the hinge base. This bone retains just the second molar (m_2). (From a tawny owl pellet)

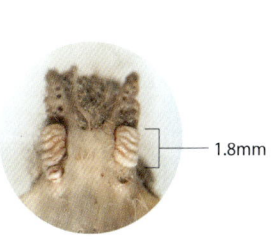
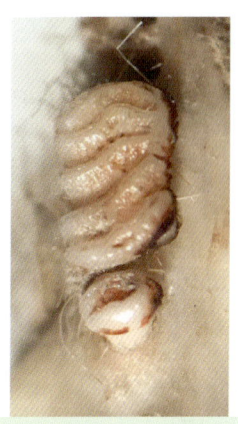

— 1.8mm

A closer view of both upper teeth rows of a dormouse followed by a magnified right-hand teeth row, showing one premolar (the smaller tooth) (p⁴) and the first molar (m¹). The bottom of each photo is the front (anterior) or nose-end of the skull. (From a tawny owl pellet)

Quick identification: Even with just a small part of the jaw and teeth remaining in a pellet, the presence of four cheek teeth with broad flattened surfaces are a quick giveaway that you are looking at the remains of a hazel dormouse. They are the colour of window putty or chewing gum. A closer look reveals low ridges running across each tooth. Skulls measure 20–23mm long (MacDonald and Barrett 1993).

Skull length range: 20–23mm (Macdonald & Barrett 1993).

Clincher: Alongside the above, if the lower jaws are present, they show a distinctive hole at the hinge base which is absent in voles, mice and rats. This base is also distinctly angled and prominent, with a diagonal straight edge. The premolar has just one root hole. The first molars (m^1 and m_1), which each have more than one root hole, are much longer across their surface than the other molars (Harris & Yalden 2008).

Edible Dormouse

The edible dormouse is a non-native rodent found in the Chilterns, in the south of England. It is three times the size of the hazel dormouse and has the same teeth arrangement (one incisor, no canines, one premolar and three molars) on each jawbone. They have fewer ridges running across each tooth compared to the hazel dormouse, and the molars are similar in size and shape to each other (Harris & Yalden 2008). Edible dormice have a much larger skull, measuring 36–44mm long (MacDonald and Barrett 1993).

Skull length range: 36–44mm (Macdonald & Barrett 1993).

[All images are life size unless otherwise indicated.]

Rabbits and hares

Rabbits and hares are common prey for many birds and are hunted by eagles, buzzards, harriers and great black-backed gulls, while dead rabbits and hares may be scavenged by corvids such as crows, magpies and ravens.

The teeth of rabbits and hares, these species collectively are known as lagomorphs, are distinctive. Rather than one incisor on each upper jaw, they have two. The second (i^2) is tiny and tucked behind the main incisors (i^1). They have no canines and a gap, the diastema, follows before the six upper cheek teeth on each jaw: three premolars (p^1 to p^3) and three molars (m^1 to m^3). On each lower jaw there is one incisor (i_1), two premolars (p_1 to p_2) and three molars (m_1 to m_3) (McDonald & Barrett 1993).

Rabbit

The rabbit is a common non-native lagomorph that is eaten by a wide range of birds of prey and scavenged by an even greater number. It is found across most of Britain and Ireland, including most islands, while absent from some Scottish islands and the Isles of Scilly. Pellets containing large, long robust bones (compared to very tiny vole or mice bones) may be from rabbits. Young rabbits are often eaten whole by great black-backed gulls and great skuas.

Two rabbit skulls, one particularly fragmented and missing the brain case (cranium).

[All images are life size unless otherwise indicated.]

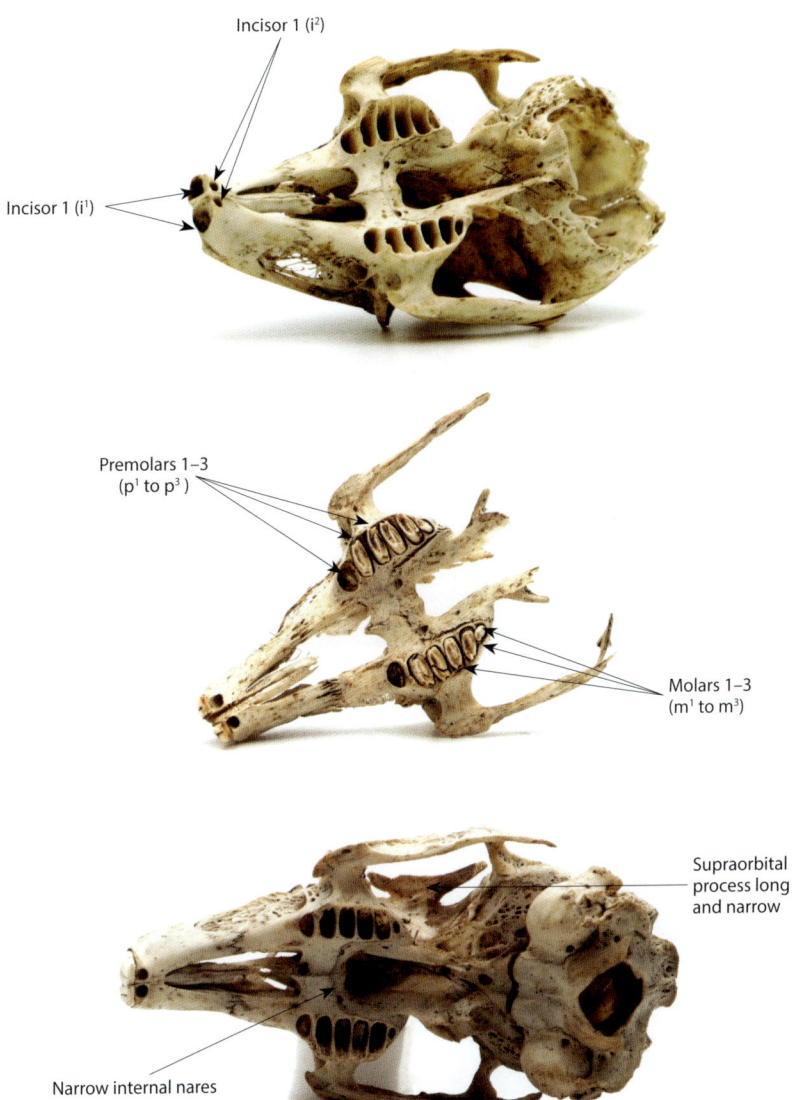

Incisor 1 (i^2)

Incisor 1 (i^1)

Premolars 1–3
(p^1 to p^3)

Molars 1–3
(m^1 to m^3)

Supraorbital
process long
and narrow

Narrow internal nares

The underside of three rabbit skulls.

[All images are life size unless otherwise indicated.]

A rabbit's humerus bone revealing the distinctive hole (foramen) at one end (right side of photo).

A rabbit's shoulder blade or scapular.

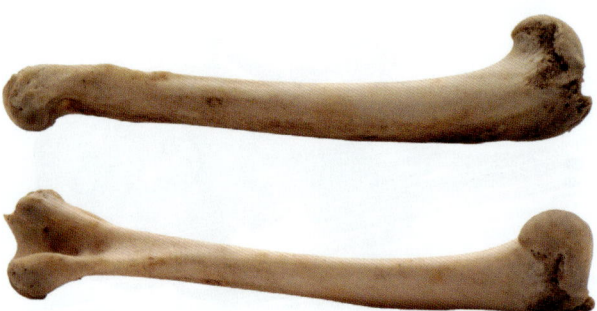

A rabbit's femur bone.

Quick identification: The skull is much bigger than that of a squirrel or rat, with a distinctive angular shape, long muzzle and gaps where intricate mesh-like bony parts within the nasal cavity are visible. Unlike rodents, rabbits have second pair of incisors (i^2) or root holes directly behind the first.

Skull length range: 68–75mm (Macdonald & Barrett 1993).

[All images are life size unless otherwise indicated.]

Clincher: If viewed from below, the internal nares (central gap between the two teeth rows) are narrow and the supraorbital process (area above the eye sockets) is thin and narrow (Harris & Yalden 2008). The skull is less than 86mm in length (Brown *et al.* 2012). The humerus bone has a distinctive hole (foramen) at one end (Yalden 2009).

Brown hare

The brown hare is found across Britain and north-west Ireland (where it was introduced). It is absent from many northern parts of Scotland and islands, although encountered on some such as many of the Inner Hebrides, the Isle of Wight, Anglesey and the Isle of Man (also introduced) (Harris & Yalden 2008; Mathews *et al.* 2018).

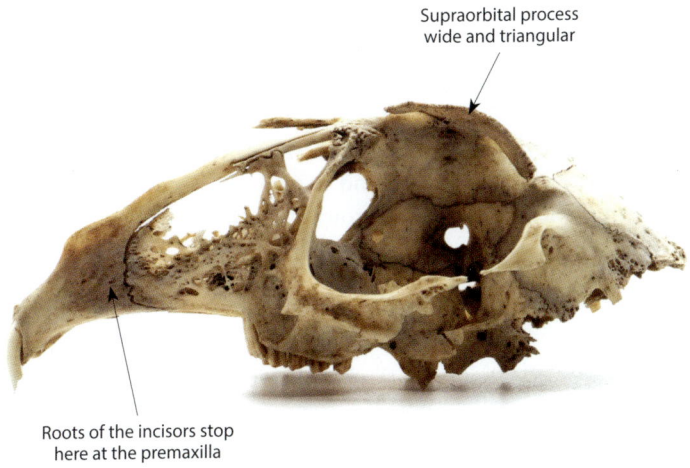

Supraorbital process wide and triangular

Roots of the incisors stop here at the premaxilla

A brown hare skull, Linz, Austria.

Incisor 1 (i²)

Incisor 1 (i¹)

Wide internal nares

The underside of brown hare skull, Linz, Austria.

[All images are life size unless otherwise indicated.]

Quick identification: The skull is similar to that of a rabbit, although larger and at first glance may be difficult to distinguish. As with the rabbit, the humerus bone has a distinctive hole (foramen) at one end.

Skull length range: 83–92mm (Macdonald & Barrett 1993).

Clincher: Although the skulls look very similar to that of a rabbit, if viewed from below the internal nares or nasal passage (central gap between the two teeth rows) are wider, the bony palate is shorter and the supraorbital process (area above the eye sockets) is broader and forms a triangular shape (more wing- or flap-like). The roots of the incisors are tightly curved and stop at the bony section called the premaxilla (small cranial bone at the tip of the upper jaw) (Harris & Yalden 2008).

Mountain hare

For some upland birds of prey, such as golden eagles, mountain hares are an important part of the diet. Their skull largely resembles that of a brown hare, and one of the main distinguishing features involves the incisors. Where the incisors disappear into the bony structure of the skull, those of the mountain hare are more gently curved and extend up and around much further, past the premaxilla into the area known as the maxilla. In the brown hare, the tooth stops short of this in the premaxilla.

The mountain hare is found across Ireland and Scotland (over 300 metres above sea level). It is also found in the Peak District and some islands including Shetland, Hoy (Orkney) and parts of the Isle of Man, Inner Hebrides and Outer Hebrides (Harris & Yalden 2008; Mathews 2018).

Skull length range: 63–80mm (Scotland) and 69–81mm (Ireland) (Macdonald & Barrett 1993).

Shrews

Shrews are an important component of the diet for many birds of prey such as owls and kestrels. The common, pygmy and water shrew are most commonly encountered across Britain, and the greater white-toothed shrew in Ireland and some of the Channel Islands. The lesser white-toothed shrew and the crowned shrew are also found on some islands (see 'Other shrews'). Depending on the age of the shrew when eaten, skulls can be variable in size. However, each species has its own diagnostic features that help confirm identification, alongside their known range in Britain and Ireland (outlined in Table 2).

When dissecting owl pellets, the smaller, narrow teardrop-shaped shrew skulls are usually obvious compared to those of voles or mice. A quick look at their jaws reveals tiny sharp insectivore teeth and, for common, pygmy, water and crowned shrews, dark-red tips. These are capped with a red-orange enamel that helps gives such tiny teeth strength and endurance. Unlike moles or hedgehogs, they lack the zygomatic arches (cheek bones) on each side of the skull (Brown *et al.* 2012). Some teeth of older shrews will be worn and not show obvious red-orange caps. On common shrews, the bumps or cusps on the lower outer incisors may also be worn and more easily resemble those of water shrews. However, they will usually still show small bumps.

The cranium is broad and round, becoming narrow where the teeth begin. After the first set of six to eight inner teeth and inner palate, the skull narrows and tapers towards the nose, ending in a pointed tip. The teeth are ideally suited for small invertebrates: those closest to the cranium are broad and sharply ridged, while the

The upper tooth row is measured after the first incisor and to the end of last molar (m^3)

The points between which the upper tooth row is measured in shrews such as this enlarged common shrew skull.

[All images are life size unless otherwise indicated.]

TABLE 2: Features of the six species of shrew found across Britain and Ireland (MacDonald & Barrett 1993; Hutterer 2005; Harris & Yalden 2008). The tooth formula shows the upper incisors and lower incisors, followed by canines, premolars and molars (i^{upper}/i_{lower}, c^{upper}/c_{lower}, p^{upper}/p_{lower}, m^{upper}/m_{lower} on one side of the skull. The total number is the grand total of teeth in the skull.

SPECIES	RANGE	TOOTH FORMULA	TEETH COLOUR	NOTABLE FEATURES	UPPER TOOTH ROW LENGTH (mm)
Common shrew	Across mainland Britain and many islands. Absent from Ireland, Shetland, Orkney, Outer Hebrides, Isle of Man, Lundy, Scilly, parts of the Inner Hebrides and the Channel Islands.	3/1, 1/1, 3/1, 3/3 = 32	Red-tipped	Lower first incisor: four bumps or cusps 5 unicuspids; 3rd smaller than 2nd	8.0–8.8
Pygmy shrew	Across Britain and Ireland. Present on most larger islands (>10km²) and many smaller islands. Absent from Shetland, Lewis, Isles of Scilly and the Channel Islands.	3/1, 1/1, 3/1, 3/3 = 32	Red-tipped	Lower first incisor: four bumps or cusps 5 unicuspids; 3rd larger than 2nd	6.2–6.6
Water shrew	Across Britain. Absent from Ireland and many small islands.	3/1, 1/1, 2/1, 3/3 = 30	Red-tipped	Lower first incisor smooth 4 unicuspids; 4th very small	8.5–9.4
Greater white-toothed shrew	Ireland and the Channel Islands (Guernsey, Alderney and Herm)	3/2, 1/0, 1/1, 3/3 = 30	White	Lower first incisor smooth 3 unicuspids; 2nd only a little bit smaller than 3rd	7.7–8.5
Lesser white-toothed shrew	Scilly and the Channel Islands (Jersey and Sark)	3/2, 1/0, 1/1, 3/3 = 30	White	Lower first incisor smooth 3 unicuspids; 2nd more obviously smaller than 3rd	7.4–8.0
Crowned shrew	Jersey (where there are no common shrews)	3/1, 1/1, 3/1, 3/3 = 32	Red-tipped	Lower first incisor: four bumps or cusps 5 unicuspids	8.0–8.8

front half are more peg-like and sharp. The incisors are longer and point outwards towards where the nose would be, particularly on the lower jaw where they may extend the jawbone by 2mm at least.

The different types of teeth on a shrew are not easily distinguished. However, the first upper incisor (i¹) is large and often hooked, followed by three to five small, single-cusped (and similar-looking) teeth known as unicuspids, of which one is a modified canine (Hutterer 2005; Harris & Yalden 2008). These are then followed by a variable number of premolars and three molars, which are much larger with more intricate (pointed) cusps and slicing edges. While the teeth are more difficult to distinguish, the length of the upper tooth row and subtle differences in the number of teeth in different sections of the upper jaw can help distinguish species (see Table 2). In the greater and lesser white-toothed shrews, the size of the second unicuspid tooth can also be useful.

Common shrew

Across their range, common shrews are frequently found in owl pellets and may even dominate their contents. They can be variable in size depending on age and may sometimes be confused with a pygmy or water shrew, depending on size. Generally, on mainland Britain, a vole-size shrew skull is likely to be from a common shrew. They are absent from Ireland, the Isle of Man, Scottish islands including Orkney, Shetland and the Outer Hebrides, Isles of Scilly and the Channel Islands (Harris & Yalden 2008).

Two common shrew skulls. (From barn owl pellets, University of Bristol)

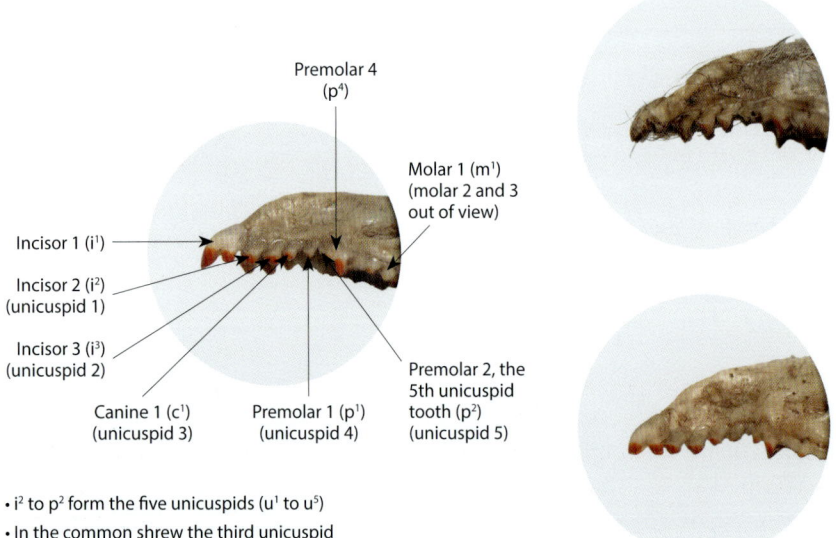

Premolar 4 (p⁴)

Molar 1 (m¹) (molar 2 and 3 out of view)

Incisor 1 (i¹)

Incisor 2 (i²) (unicuspid 1)

Incisor 3 (i³) (unicuspid 2)

Canine 1 (c¹) (unicuspid 3)

Premolar 1 (p¹) (unicuspid 4)

Premolar 2, the 5th unicuspid tooth (p²) (unicuspid 5)

- i² to p² form the five unicuspids (u¹ to u⁵)
- In the common shrew the third unicuspid (c¹) is smaller than the second (i³)

The upper teeth of three common shrews revealing the double-cusped first incisor and the five unicuspid teeth (the 5th is very small and can be hard to see) (×20 magnification). (From barn owl pellets, University of Bristol)

The underside of a common shrew skull. (From a barn owl pellet, University of Bristol)

[All images are life size unless otherwise indicated.]

A selection of common shrew lower-left jawbones, revealing the four bumps or cusps on the first incisor. The outer or front two cusps are close to each other and almost disappear in older individuals. (From barn owl pellets, University of Bristol)

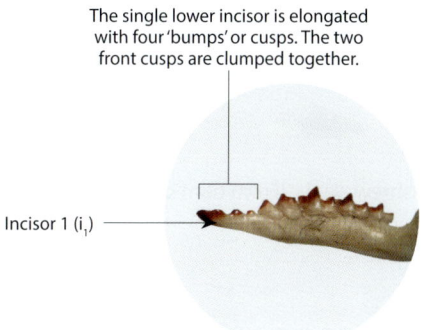

The single lower incisor is elongated with four 'bumps' or cusps. The two front cusps are clumped together.

Incisor 1 (i_1)

A closer view of the lower-left jawbone of a common shrew (×20 magnification). (From a barn owl pellet, University of Bristol)

Quick identification: Common larger teardrop-shaped shrew skull with red-tipped teeth in pellets.

Skull length range: 18–20mm (Macdonald & Barrett 1993).

Clincher: Five upper unicuspid teeth. Four small bumps or cusps on the lower first incisor. The two towards the tip are close together and almost merge. In older animals they may be difficult to tell as separate. Upper tooth row 8.0–8.8mm (Harris & Yalden 2008). To distinguish between young common shrews and full-size pygmy shrews, the second unicuspid tooth is larger than the third unicuspid tooth in common shrews and smaller than the third unicuspid tooth in pygmy shrews.

[All images are life size unless otherwise indicated.]

Pygmy shrew

The skulls of pygmy shrews are tiny and can be distinguished from the common shrew both by their much smaller size (although young common shrews may be small) and differences in the size of the third unicuspid compared to the second (see below).

The pygmy shrew is found across Britain and Ireland and on many islands, although absent from Shetland, Lewis, Isles of Scilly and the Channel Islands (Harris & Yalden 2008).

A selection of pygmy shrew skulls. (From barn owl pellets, University of Bristol)

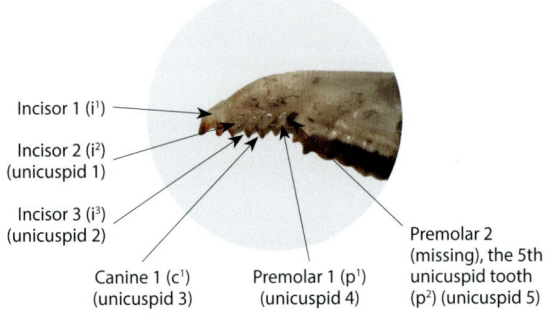

Incisor 1 (i^1)

Incisor 2 (i^2) (unicuspid 1)

Incisor 3 (i^3) (unicuspid 2)

Canine 1 (c^1) (unicuspid 3)

Premolar 1 (p^1) (unicuspid 4)

Premolar 2 (missing), the 5th unicuspid tooth (p^2) (unicuspid 5)

• i^2 to p^2 form the five unicuspids (u^1 to u^5)
• In the pygmy shrew the third unicuspid (c^1) is larger than the second (i^3)

The upper teeth of a pygmy shrew, revealing the double-cusped first incisor and the position of the five unicuspid teeth (×20 magnification). The fifth unicuspid tooth is missing (fallen out) from this skull. The third unicuspid tooth is larger than the second. (From a barn owl pellet, University of Bristol)

Another pygmy shrew with the full complement of upper teeth (×20 magnification) (cat kill).

[All images are life size unless otherwise indicated.]

The underside of a pygmy shrew skull. (From a barn owl pellet, University of Bristol)

A selection of pygmy shrew lower-left jawbones, revealing the four bumps or cusps on the first incisor. Each cusp is evenly spaced rather than the outer two cusps clumped as in the common shrew. (From barn owl pellets, University of Bristol)

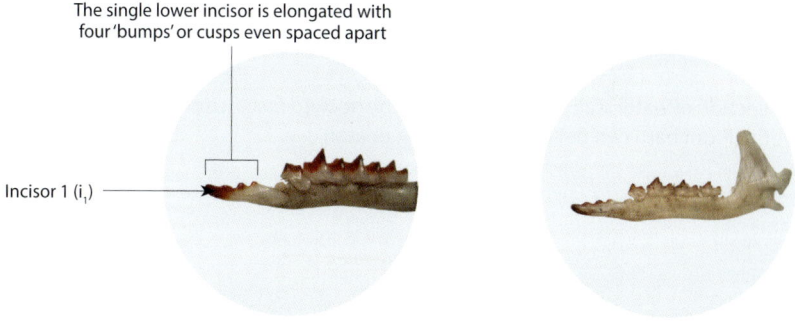

The single lower incisor is elongated with four 'bumps' or cusps even spaced apart

Incisor 1 (i$_1$)

Closer views of the lower-left jawbones of two pygmy shrews (×20 magnification). (From barn owl pellets, University of Bristol)

Quick identification: Tiny and delicate teardrop-shaped shrew skull with red-tipped teeth in pellets. Half to two thirds of the size of common shrew.

Skull length range: 14.8–16.7mm (Macdonald & Barrett 1993).

Clincher: Five upper unicuspid teeth with the third larger than the second. Four small bumps or cusps on the lower first incisor evenly spaced.

[All images are life size unless otherwise indicated.]

Water shrew

At first glance, water shrew skulls resemble those of common shrews, although they are much less common in pellets. However, with experience, their larger size and broader palate, along with longer, kinked upper incisors help give away their presence. The long incisors on the lower jaw lack any bumps or cusps.

The water shrew is found across Britain, particularly in the south and central areas. It is absent from Ireland and many islands. It is, however, found on islands such as Anglesey, Isle of Wight and Hoy in Orkney (Harris & Yalden 2008).

A selection of water shrew skulls, revealing the hooked first incisors and four unicuspid teeth. (From barn owl pellets, University of Bristol)

4.9mm

The undersides of water shrew skulls, revealing the intricate pattern of cutting edges. (From barn owl pellets, University of Bristol)

[All images are life size unless otherwise indicated.]

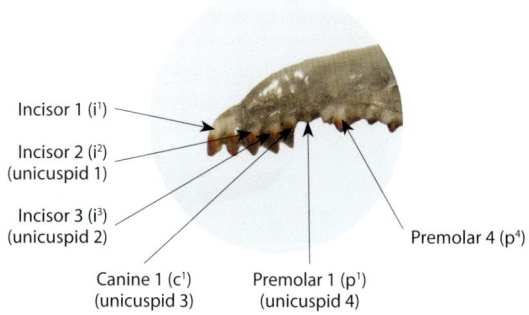

Incisor 1 (i¹)

Incisor 2 (i²)
(unicuspid 1)

Incisor 3 (i³)
(unicuspid 2)

Premolar 4 (p⁴)

Canine 1 (c¹)
(unicuspid 3)

Premolar 1 (p¹)
(unicuspid 4)

• i² to p¹ form the four unicuspids (u¹ to u⁴)

A comparison of the upper teeth of water shrews, revealing the hooked first incisors and the four unicuspid teeth (×20 magnification). (From barn owl pellets, University of Bristol)

The front portion of a water shrew skull (upper jaw), with the incisor and unicuspid teeth missing. (From a barn owl pellet, University of Bristol)

2.8mm

The lower jawbones of water shrews, revealing the smooth first incisor and absence of any bumps or cusps. (From barn owl pellets, University of Bristol)

[All images are life size unless otherwise indicated.]

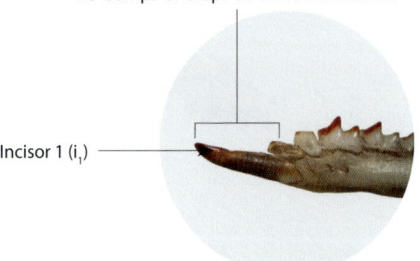

Unlike common or pygmy shrews there are
no 'bumps' or cusps on the lower incisors.

Incisor 1 (i₁)

A closer view of the lower-left jawbone of a water shrew (×20 magnification). (From a barn owl pellet, University of Bristol)

Quick identification: Similar to common shrew, although larger and broader or more bulbous with a convex forehead (Yalden 2009). The water shrew has longer, more pronounced lower incisors (a millimetre or more in length), which look like miniature daggers. The upper incisors of water shrews are more pronounced than those for common shrews, with a distinct kink.

Skull length range: 19–22mm (Macdonald & Barrett 1993).

Clincher: Four upper unicuspid teeth with fourth smaller than the first three. A smooth first lower incisor with just a gentle, subtle bump or lobe (rather than any distinctive cusps). The inner cusp of the first upper incisor is proportionally smaller compared to those in the common and pygmy shrews. Upper tooth row 8.5–9.4mm (Harris & Yalden 2008).

Other shrews

The Republic of Ireland and some smaller islands in the south of Britain – such as Scilly, Jersey and Guernsey – host different shrew species not found on the mainland. These are the greater white-toothed shrew, the lesser white-toothed shrew and the crowned shrew (also known as Millet's shrew or the French shrew). Both the white-toothed shrews, as their names suggest, have white teeth and live in different locations to each other, making identification easier. The crowned shrew is almost identical to the common shrew, but it lives only on Jersey where the latter does not occur.

Greater white-toothed shrew

Greater white-toothed shrew skulls most resemble those of pygmy shrews in shape, although they are large as those of water shrews. Unlike common, pygmy and water shrews, they lack the red-orange tips to their teeth. They occur in Ireland and on the Channel Islands of Guernsey, Alderney and Herm, where they are frequently encountered in barn owl and hen harrier pellets (Harris & Yalden 2008). This species has also been discovered in the north-east of England since 2015, their origin unknown.

A selection of greater white-toothed shrew skulls, revealing their white teeth and three unicuspid teeth. (From barn owl pellets, collected by Alan McCarthy)

The underside of a greater white-toothed shrew skull. (From a barn owl pellet, collected by Alan McCarthy)

The underside of a greater white-toothed shrew skull and the front view of the same skull, revealing its pair of first white incisors and chunky, slicing white premolars. (From a barn owl pellet, collected by Alan McCarthy)

A selection of greater white-toothed shrew lower jawbones, revealing their smooth white first incisors and the absence of any cusps. (From barn owl pellets, collected by Alan McCarthy)

[All images are life size unless otherwise indicated.]

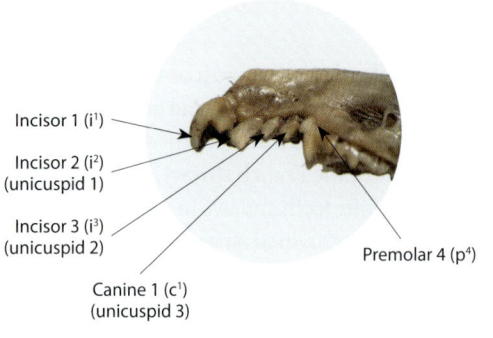

Incisor 1 (i¹)

Incisor 2 (i²)
(unicuspid 1)

Incisor 3 (i³)
(unicuspid 2)

Canine 1 (c¹)
(unicuspid 3)

Premolar 4 (p⁴)

- i² to c¹ form the three unicuspids (u¹ to u³)
- Distinguished from lesser white toothed shrew by the second unicuspid (i³) only being a little smaller than the third (c¹).

A comparison of the upper teeth of two greater white-toothed shrews (×20 magnification). (From barn owl pellets, collected by Alan McCarthy)

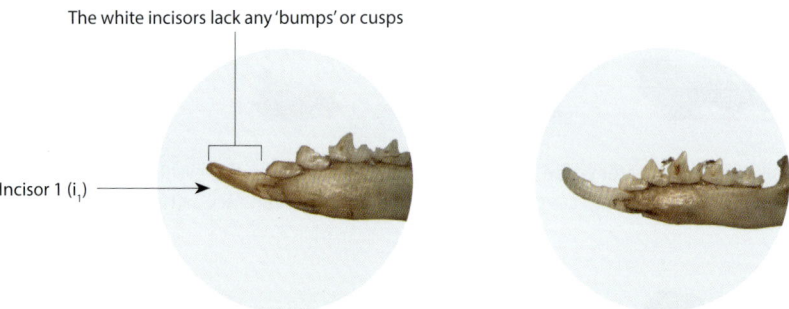

The white incisors lack any 'bumps' or cusps

Incisor 1 (i₁)

A closer view of the lower-left jawbones of greater white-toothed shrews (×20 magnification). (From barn owl pellets, collected by Alan McCarthy)

Quick identification: The white teeth, size and location/range all help with quick identification.

Clincher: Lower first incisor smooth. Three upper unicuspid teeth, with the second only a little bit smaller than the third. Upper tooth row 7.7–8.5mm (Harris & Yalden 2008). Maximum length of the skull is generally over 18.5mm (range 18–20.4mm), and the lower jaw is more than 10mm long (Macdonald & Barrett 1993, Brown *et al.* 2012).

[All images are life size unless otherwise indicated.]

Lesser white-toothed shrew

The lesser white-toothed shrew is found on the Scilly Isles and on the Channel Islands of Jersey and Sark (Harris & Yalden 2008). Its skull is very similar to that of the greater white-toothed shrew. A closer look at the teeth reveals that the second unicuspid tooth is more obviously smaller than the third. In the greater white-toothed shrew, the size difference is only very small. While both species are found together on mainland Europe, the two species occupy different islands in Britain.

A selection of lesser white-toothed shrew skulls, revealing the white teeth and three unicuspid teeth, Molène archipelago, Brittany, France. (From barn owl pellets, Bernard Cadiou)

Another lesser white-toothed shrew skull. (From a barn owl pellet, collected by the Jersey Barn Owl Conservation Trust and Ian Buxton)

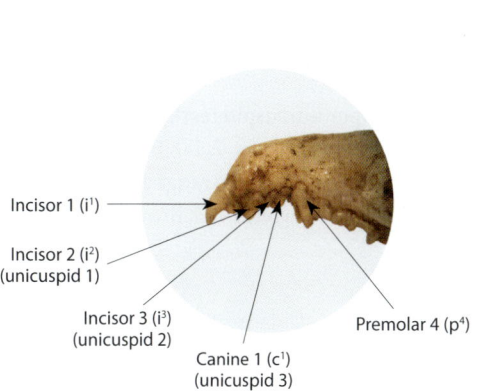

Incisor 1 (i¹)
Incisor 2 (i²)
(unicuspid 1)
Incisor 3 (i³)
(unicuspid 2)
Canine 1 (c¹)
(unicuspid 3)
Premolar 4 (p⁴)

• i² to c¹ form the three unicuspids (u¹ to u³)
• Distinguished from greater white toothed shrew by the second unicuspid (i³) being more obviously smaller than the third (c¹).

A comparison of the upper teeth of three lesser white-toothed shrews, revealing the white teeth and three unicuspid teeth (×20 magnification). Molène archipelago, Brittany, France. (From barn owl pellets, Bernard Cadiou)

[All images are life size unless otherwise indicated.]

The lower left jawbone of a lesser white-toothed shrews. Molène archipelago, Brittany, France (From a barn owl pellet, Bernard Cadiou)

Another lower left jawbone of a lesser white-toothed shrew (From a barn owl pellet, collected by the Jersey Barn Owl Conservation Trust and Ian Buxton)

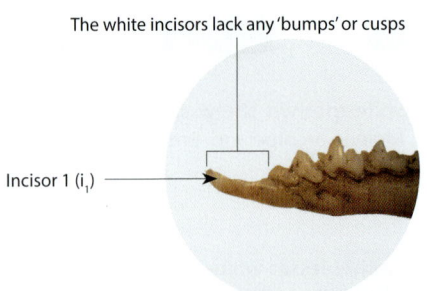

The white incisors lack any 'bumps' or cusps

Incisor 1 (i₁)

A closer view of the lower-left jawbone of a lesser white-toothed shrew (×20 magnification). Molène archipelago, Brittany, France. (From a barn owl pellet, Bernard Cadiou)

Quick identification: The white teeth, size and location and range all help with quick identification.

Clincher: Lower first incisor smooth. Three upper unicuspid teeth, with the second more obviously smaller than the third. Upper tooth row 7.4–8.0mm (Harris & Yalden 2008). Maximum length of the skull less than 18.5mm (range 15.6–18.3mm), and the lower jaw is less than 10mm (Macdonald & Barrett 1993, Brown *et al.* 2012).

[All images are life size unless otherwise indicated.]

Crowned shrew

The crowned shrew, also known as the Jersey shrew or Millet's shrew, resembles the common shrew and subtle differences separate their skulls and lower jaws. Fortunately, the two species do not overlap their range in Britain. The crowned shrew is found just on Jersey. It also occupies western France where the common shrew does not occur (Harris & Yalden 2008).

Two crowned shrew skulls, Western Brittany, France. (Collected by Bernard Cadiou)

Another crowned shrew skull. (From a barn owl pellet, collected by the Jersey Barn Owl Conservation Trust and Ian Buxton)

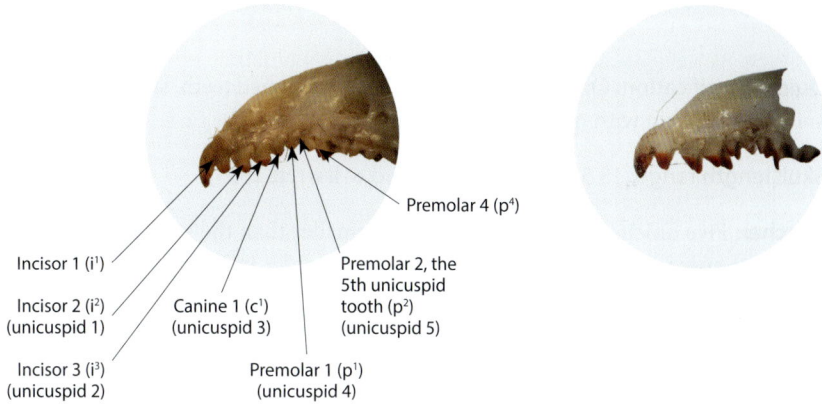

Premolar 4 (p⁴)

Incisor 1 (i¹)

Incisor 2 (i²)
(unicuspid 1)

Canine 1 (c¹)
(unicuspid 3)

Premolar 2, the
5th unicuspid
tooth (p²)
(unicuspid 5)

Incisor 3 (i³)
(unicuspid 2)

Premolar 1 (p¹)
(unicuspid 4)

- i² to p² form the five unicuspids (u¹ to u⁵)
- In the crowned shrew (as in the common shrew, the third unicuspid (c¹) is smaller than the second (i³)

A closer view of the upper-left teeth of two crowned shrews, revealing the double-cusped first incisor and the five unicuspid teeth (×20 magnification). Western Brittany, France. (Bernard Cadiou)

The lower-left jawbone of a crowned shrew, revealing the four bumps or cusps on the first incisor, Molène archipelago, Brittany, France. (From a barn owl pellet, Bernard Cadiou)

[All images are life size unless otherwise indicated.]

Another lower-left jawbone of a crowned shrew. (From a barn owl pellet, collected by the Jersey Barn Owl Conservation Trust and Ian Buxton)

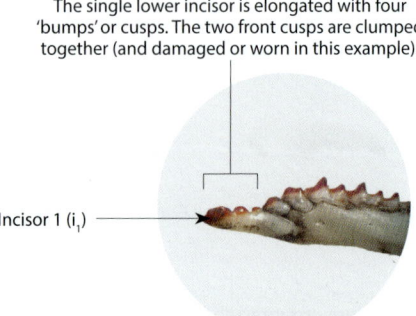

The single lower incisor is elongated with four 'bumps' or cusps. The two front cusps are clumped together (and damaged or worn in this example).

Incisor 1 (i_1)

A closer view of the lower-left jawbone of a crowned shrew (×20 magnification). Molène archipelago, Brittany, France. (From a barn owl pellet, Bernard Cadiou)

Quick identification: Only found on Jersey. The red-tipped teeth, size and location and range all help with quick identification.

Skull length range: 18.5–20mm (Macdonald & Barrett 1993).

Clincher: Five unicuspid teeth with the third smaller than the second. Four small bumps or cusps on the lower first incisor, worn in older individuals.

[All images are life size unless otherwise indicated.]

Bats

Bats feature – to varying degrees – in a wide range of different bird diets, from black-headed gull to rook and birds of prey such as peregrine, kestrel, hobby, sparrowhawk, tawny owl, little owl and barn owl (Speakman 1991; Mikula *et al.* 2016). While for many predator species bats are uncommon or rare in their pellets, for birds that eat them more regularly, such as hobbies, their fur and bones are more frequent or common.

Bat skulls are full of tiny sharp white teeth – on the noctule bat, for example, the premolars and molars are sharp with multiple slicing surfaces. The tip of the nose area is more open and wider in bat skulls than in other small mammals. Some skulls resemble the shape of weasel/stoat skulls, while others, such as the long-eared bats and pipistrelles, have a steeply sloping cranium that is raised above the jaws. The surface of a bat's skull has various areas for strong muscle attachment; rather than being smooth and rounded, the surface is more complex and has more crater-like indentations compared to the skulls of rodents.

Noctule bat

A noctule bat skull revealing its angular profile. The large skull has 34 teeth in total. Each upper jaw has two incisors, one canine, two premolars and three molars. The first upper premolar, p^1, is tiny and tucked behind the incisor and the next premolar (Harris & Yalden 2008). Each lower jaw has three incisors, one canine, two premolars and three molars (Brown *et al.* 2012). (Bristol Museum & Art Gallery)

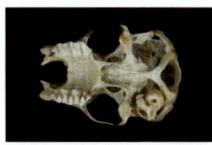

The underside of a noctule bat skull. (Bristol Museum & Art Gallery)

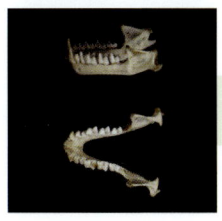

A noctule bat lower jawbone, from the side and from above. (Bristol Museum & Art Gallery)

Skull length range: 17.4–19.9mm (Macdonald & Barrett 1993).

[All images are life size unless otherwise indicated.]

Brown long-eared bat

A brown long-eared bat skull measures 13–15.5mm long (over 17mm in grey-long eared (Brown *et al.* 2012)). There are 36 teeth in total. Each upper jaw has two incisors, one canine, two premolars and three molars. Each lower jaw has three incisors, one canine, three premolars and three molars. (Bristol Museum & Art Gallery)

The underside of a long-eared bat skull. (Bristol Museum & Art Gallery)

A pair of lower jawbones from a long-eared bat. (Bristol Museum & Art Gallery)

The leaf-shaped shoulder blade or scapular from a long-eared bat. (Bristol Museum & Art Gallery)

The tail and part of the pelvic girdle or sacrum from a long-eared bat. (Bristol Museum & Art Gallery)

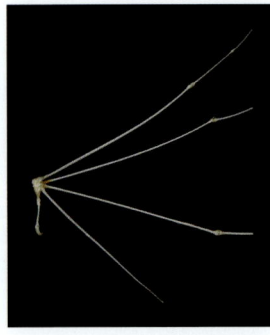

The wing bones from a long-eared bat. (Bristol Museum & Art Gallery)

[All images are life size unless otherwise indicated.]

Lesser horseshoe bat

Lesser horseshoe bat skull, which is one third the size of a greater horseshoe bat skull, measuring 13.5–15.2mm long (Harris & Yalden 2008). The brain case or cranium is partially collapsed in this example. Their skulls would usually have two bulbous or bony lumps on top behind the nasal cavity. The skull has 32 teeth in total. Each upper jaw has one incisor, one canine, two premolars and three molars. Each lower jaw has two incisors, one canine, three premolars and three molars (Brown *et al.* 2012). (Bristol Museum & Art Gallery)

The front (nose) view of lesser horseshoe bat skull. The brain case or cranium is partially collapsed. (Bristol Museum & Art Gallery)

Skull length range: 13.4–16mm (Macdonald & Barrett 1993).

Pipistrelle bats

The first upper premolar is tiny and slightly tucked behind the large canine

The skull of either a common or soprano pipistrelle bat alongside a closer view (×20 magnification) of the teeth (Bristol Museum & Art Gallery). The two species both have 34 teeth in total. The first upper premolar is tiny and slightly tucked behind the large canine. Incisors vary in size. Each upper jaw has two incisors, one canine, two premolars and three molars. Each lower jaw has three incisors, one canine, two premolars and three molars (Brown *et al.* 2012). Distinguishing between common and soprano pipistrelle remains is mostly done through detailed analysis of skull micro-measurements (Barlow 1997). Common pipistrelle generally has a larger skull, longer upper canines and longer lower jaws. Common pipistrelle skulls are 10.8–12.2mm long and soprano pipistrelle skulls 10.9–11.9mm long (Harris & Yalden 2008).

[All images are life size unless otherwise indicated.]

Other resources to help identify bat remains

The following resources also help with the identification of bat remains found in bird pellets.

Illustrations and teeth formula for bat species:
*   Brown, R.W., Lawrence, M.J. and Pope, J. 2012. *Animals Tracks, Trails & Signs*. Octopus Publishing Group Ltd, London.

Hair patterns for bat hair:
*   Dietz, C. and Kiefer, A. 2016. *Bats of Britain and Europe*. Bloomsbury Natural History, London.

Details on teeth:
*   Harris, S. and Yalden, D.W. 2008. *Mammals of the British Isles. Handbook 4th Edition*. The Mammal Society, Southampton.

Out of print and may be available through second-hand online shops or academic libraries:
*   Yalden, D.W. 1985. *The identification of British Bats*. Occasional Publication No. 5. The Mammal Society, London.

Measurements for skull and other anatomy:
*   Stebbings, H.N., Yalden, D.W. and Herman, J.S. 2007. *Which Bat Is It?* The Mammal Society, London.

Mole

Moles come up from underground and move across the surface when dispersing or looking for a mate. This makes them vulnerable to predation both from diurnal predators, such as gulls and buzzards, and nocturnal predators, such as owls. They are insectivores with skulls narrower and more elongated than in a mouse or vole. They more closely resemble an overstretched shrew skull, although are much larger (over 25mm) (Brown *et al.* 2012).

Moles are found across Britain but are absent from Ireland, the Isle of Man and many Scottish islands such as Orkney, Shetland and the Outer Hebrides (Harris *et al.* 2008; Mathews *et al.* 2018).

A mole skull. (From a barn owl pellet, University of Bristol)

Another mole skull. (From a barn owl pellet, University of Bristol)

A mole lower left jawbone. (From a barn owl pellet, University of Bristol)

Another mole lower left jawbone with the majority of teeth missing. (From a barn owl pellet, University of Bristol)

A mole front leg with the bones and digging claws all still intact. (From a barn owl pellet, University of Bristol)

A complete mole pelvis bone with tail bones attached (seen on right-hand side). (From a barn owl pellet, University of Bristol)

A slightly incomplete mole pelvis bone, viewed from above. (From a barn owl pellet, University of Bristol)

[All images are life size unless otherwise indicated.]

The distinctively shaped humerus of a mole. This provides the ultimate anchorage for muscles and articulation to allow a mole to do all its digging. (From a barn owl pellet, University of Bristol)

The femur from a mole. (From a barn owl pellet, University of Bristol)

Quick identification: The teeth, as with other insectivores, are very similar to each other and mostly small and pointed. They lack the red-orange tips of shrews. The shrew-like skull is much larger than that of a water shrew or greater white-toothed shrew. Even if the brain case is missing, the remaining parts are smooth, elongated and tapering.

Skull length range: 32.5–37mm (Macdonald & Barrett 1993).

Clincher: Moles have a thin cheek bone (zygomatic arch) which is absent in shrews (Brown *et al.* 2012). Moles have the greatest number of teeth (44) of any mammal in Britain and Ireland. On each of the upper and lower jaws they have three incisors, one canine, four premolars and three molars. The incisors look very similar to each other and on the lower jaw the canine resembles the incisors. The first lower premolars are then larger and suited for slicing and crunching earthworms, while the remaining premolars and molars are smaller and similar to each other (MacDonald & Barrett 1993; Harris & Yalden 2008). Often a mole's distinctive shovel-like front feet remain intact and are an easy giveaway. The pelvis is long and very narrow, and the humerus is short and oblong with intricate curves and pits that allow for strong muscle attachment.

[All images are life size unless otherwise indicated.]

Hedgehog

Hedgehogs may appear in the diet of some birds, particularly scavengers such as crows, magpies, ravens, buzzards and red kites. They are most likely to be eating hedgehogs that have been killed by cars or other animals such as badgers and foxes. Eagle owls and golden eagles will also eat many hedgehogs. The spines may be swallowed and therefore provide an obvious identification to hedgehog prey. Hedgehogs are found across Britain and Ireland, including many islands, mostly where grassland and woodland habitats are situated close together (Harris *et al.* 2008; Mathews *et al.* 2018).

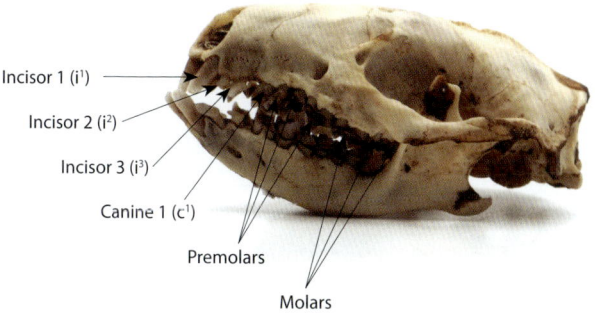

Incisor 1 (i^1)
Incisor 2 (i^2)
Incisor 3 (i^3)
Canine 1 (c^1)
Premolars
Molars

A hedgehog skull.

The underside of a hedgehog skull. (Bristol Museum & Art Gallery)

[All images are life size unless otherwise indicated.]

The underside of another hedgehog skull. (Bristol Museum & Art Gallery)

A pair of hedgehog lower jawbones. (Bristol Museum & Art Gallery)

Quick identification: The skull is bulky and robust with a shallow ridge along its top (the sagittal crest) which is for muscle attachment. The cheek bone is obvious. Ten upper and eight lower teeth on each side with front teeth cone-shaped or peg-like.

Skull length range: 55–60mm (Macdonald & Barrett 1993).

Clincher: The upper jaws each contain three incisors, one canine, three premolars and three molars. Each lower jawbone contains two incisors, one canine, two premolars and three molars. The front incisors are pronounced and obvious (while the canines are more subtle). The remaining incisors, canines and first premolars are small and peg-like. The rear teeth (one of the premolars and molars) have become more specialised and ridged (like miniature mountain ridges), with broad, crushing surfaces. The lower jaw is over 30mm long (Brown *et al.* 1995; Harris & Yalden 2008).

Stoat and weasel

Stoats and weasels may feature in the diet of birds, either having been hunted or found as roadkill. A range of birds of prey feed on these mustelids, including eagles, buzzards and owls.

Stoats are found across Britain and Ireland, including many islands such as Anglesey, Isle of Wight, Isle of Sheppey and Isle of Man and those in Scotland such as Orkney, Shetland and Mull (Harris & Yalden 2008; Mathews *et al.* 2018). Weasels occur on mainland Britain but are absent from Ireland. They are also present on some islands such as the Isle of Wight, Isle of Sheppey, Isle of Bute and Anglesey (Harris & Yalden 2008; Mathews 2018).

A stoat skull. (Bristol Museum & Art Gallery)

A weasel skull. (Bristol Museum & Art Gallery)

A pair of stoat lower jawbones. (Bristol Museum & Art Gallery)

A pair of weasel lower jawbones. (Bristol Museum & Art Gallery)

[All images are life size unless otherwise indicated.]

A close-up view of a young weasel's left jawbone.

Quick identification: Stoat and weasel skulls are elongated like a jellybean, with the eye sockets and teeth all clustered towards the very front of the skull, a blunt nose hole and a shallow sagittal crest running along the top. They have stout, broad lower jaws which are gently curved along their length and distinctly broad and angular where the muscles attach. The teeth are sharp, with pointed cusps.

Clincher: Stoats and weasels have smaller skulls and fewer molars than hedgehogs. Each upper jaw has three tiny incisors, one long sharp canine, three premolars and one molar. This is repeated on each lower jaw, plus an additional molar. Their premolars and molars are distinctive, with sharp, chomping edges and surfaces. The fourth premolar (p^4) on the upper jaw and the first molar (m_1) on the lower jaw form slicing, shearing teeth known as carnassials (these are also present in other carnivores) (MacDonald & Barrett 1993; Harris & Yalden 2008).

The two species are generally distinguishable on skull size, although larger male weasels may overlap with smaller female stoats. Stoat skulls are generally longer than 42mm (range 38.8–49.6mm), while weasel skulls range 32–44mm long, hence the potential overlap (Macdonald & Barrett 1993, Harris & Yalden 2008). The length of the lower jaw in stoats is more than 21mm, and in weasels ranges 15–22mm (Harris & Yalden 2008).

12 IDENTIFYING OTHER SMALL ANIMAL PARTS

Amphibians

Frogs, toads and newts may all be eaten by birds of prey, particularly owls, buzzards and honey-buzzards, and other birds such as carrion crows, ducks, herons, egrets, bitterns and cormorants. Wet nights that coincide with movements of amphibians under the cover of darkness provide ideal feeding opportunities for nocturnal birds of prey. Other amphibians may be taken while resting and breeding in pools and lakes.

The bones of amphibians, in particular frogs (which are more palatable than toads), often stand out as something different when encountered in a pellet. At first glance, they may look like the longer wing or leg bones of a bird and yet not quite fit the correct shape. Unlike mammal or bird skulls, the skull of an amphibian tends to break down into its small individual parts rather than stay intact. The skin may also be left in the pellet and when wet will be soft and slippery (yet tough), or more crisp and brittle when dry – a little like the texture of seaweed!

One resource to help with identifying particular bones of this group is a manual produced as part of Dr Chris Gleed-Owen's doctoral thesis, which can be found through search engines online. The manual is for the identification of fossil herpetofaunal remains and forms Chapter 5 in the thesis: 'Quaternary herpetofaunas of the British Isles: Taxonomic descriptions, palaeoenvironmental reconstructions, and biostratigraphic implications' (Coventry University, March 1998).

Chris offers tips on identifying reptile and amphibian remains from bird pellets: 'Frog/toad limb bones are recognisable as being hollow tubes, much like bird bones, but with flared ends. One very recognisable bone is the tibiofibula, the fused pair of lower-leg bones that I always think look like a double-barrelled shotgun. Frogs have longer narrower leg bones, toads have shorter broader ones. The arms are similar in frogs and toads. Vertebrae and head bones are diagnostic, but tricky for the non-expert.'

These photos below provide a guide to the general shape and form of bones to look for when frogs and toads are found in pellets. For more detailed identification and analyses, do refer to Chris's work.

Note: All images are shown life size unless a scale bar or a note in the photo caption indicates otherwise.

Common Frog

Common frogs are often found in the diet of birds of prey and sometimes corvids, such as crows. They are hunted at night during wet weather or picked up as roadkill. They are found across Britain and Ireland.

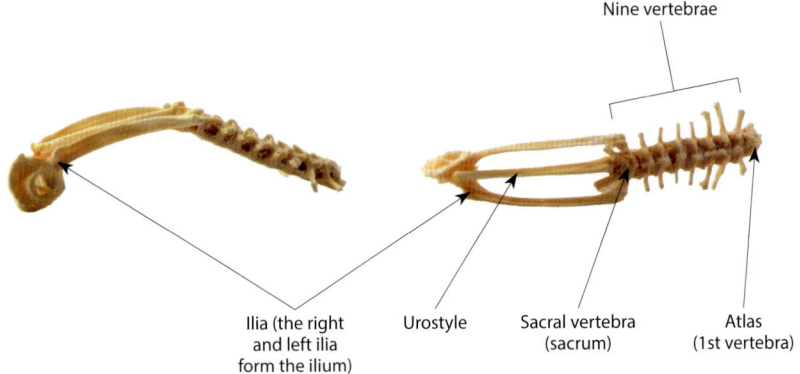

Nine vertebrae

Ilia (the right and left ilia form the ilium)

Urostyle

Sacral vertebra (sacrum)

Atlas (1st vertebra)

The pelvic girdle and vertebrae of a common frog viewed from the side and from above. Frogs and toads have nine vertebrae. (Bristol Museum & Art Gallery)

The pectoral girdle of a common frog, connecting the front leg bones to the rest of the skeleton. (Bristol Museum & Art Gallery)

Radius

Ulna

The front leg of a common frog. (Bristol Museum & Art Gallery)

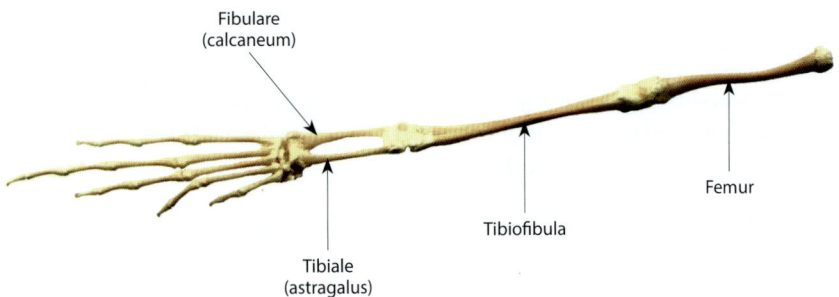

Fibulare (calcaneum)

Tibiale (astragalus)

Tibiofibula

Femur

The rear leg and foot bones of a common frog. (Bristol Museum & Art Gallery)

[All images are life size unless otherwise indicated.]

The humerus of a common frog. Gently twisted and like the arm of some elegant piece of furniture. (Bristol Museum & Art Gallery)

The front view of a frog skull. (Bristol Museum & Art Gallery)

The underside of a frog skull. (Bristol Museum & Art Gallery)

A pair of lower jawbones from a frog, which have teeth – unlike those of common toads which do not. (Bristol Museum & Art Gallery)

Common Toad

Common toads also feature in the diet of birds, although they may be less palatable than frogs due to their poisonous secretions. Care is needed here, as they can be misidentified as frogs by those dissecting pellets. Like frogs, they are most likely hunted at night during wet weather or picked up as roadkill. Toads are found across Britain but are absent from Ireland. They have shorter limbs than frogs and lack teeth.

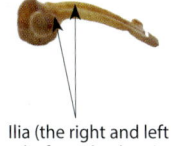

The pelvic girdle of a common toad, connecting the back leg bones to the rest of the skeleton. (Bristol Museum & Art Gallery)

Ilia (the right and left ilia form the ilium)

The pectoral girdle of a common toad, connecting the front leg bones to the rest of the skeleton. (Bristol Museum & Art Gallery)

[All images are life size unless otherwise indicated.]

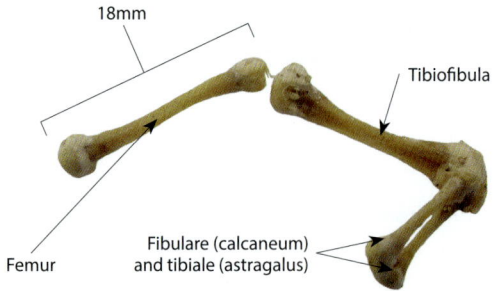

18mm

Tibiofibula

Femur

Fibulare (calcaneum)
and tibiale (astragalus)

The rear leg bones of a common toad. (Bristol Museum & Art Gallery)

Humerus

Radius

Radioulna
(the radius and
ulna together)

Ulna

The front leg bones of a common toad. (Bristol Museum & Art Gallery)

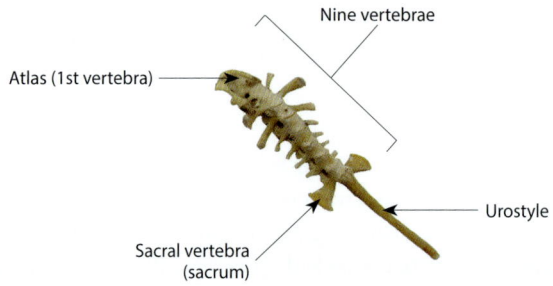

Nine vertebrae

Atlas (1st vertebra)

Urostyle

Sacral vertebra
(sacrum)

The sacrum and nine vertebrae of a common toad. A tenth rod-like vertebra, the urostyle, articulates with the sacrum and sits between the ilia (see pelvic girdle). (Bristol Museum & Art Gallery)

A pair of lower jawbones from a common toad, which lack teeth – unlike a frog which has teeth. (Bristol Museum & Art Gallery)

[All images are life size unless otherwise indicated.]

Newts

As is the case with frogs and toads, newts are potentially easy to hunt, particularly during the spring when they are moving through the landscape to find ponds and lakes to breed in. Their soft bodies and thin bones probably mean many newt skeletons are digested or go unidentified in pellets. In Britain there are three species of newt: the common or smooth newt, the palmate newt and the great crested newt. In Ireland only the smooth newt occurs.

7.5mm

The skeleton of a common or palmate newt. (Bristol Museum & Art Gallery)

4mm (femur)

The pelvis and legs of a common or palmate newt. (Bristol Museum & Art Gallery)

Reptiles

As with amphibians, the numerous bones that make up the skulls of snakes and lizards may fall apart when digested or the skulls may be damaged when the animal's head is crushed by the predator to kill it. The jawbones of snakes have tiny, backward-pointing teeth. Often the giveaway will be remnants of their scaly skin and, for snakes, multiple ribs and vertebrae.

Again, Dr Chris Gleed-Owen explains: 'Newt and lizard bones are relatively poorly preserved. Snake vertebrae preserve better, and are pretty recognisable, with lots of bits jutting out. Slow-worms (legless lizards) have one of the most recognisable features: their skin contains thousands of ossified scale cores called osteoderms. They are small roundish plates about 1–2mm across, slightly curved, and have a rugose (knobbly) upper surface, and a smooth lower surface.'

In Ireland there are fewer reptiles to be predated on, just the viviparous or common lizard and the slow worm (introduced on the Burren, County Clare).

Snakes

Different parts of a bird of prey pellet containing the scaly skin and multiple ribs and vertebrae of a snake. The skin looks papery. Near Embalse de Arroyo Bremudo, Spain.

Snake bones

These images show different parts of the above pellet under a microscope, revealing different parts of the snake that was eaten.

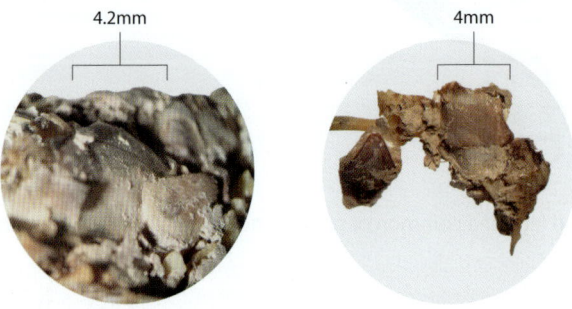

4.2mm   4mm

Scales.

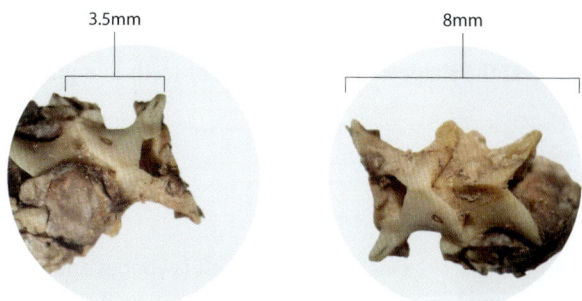

3.5mm

8mm

Vertebrae.

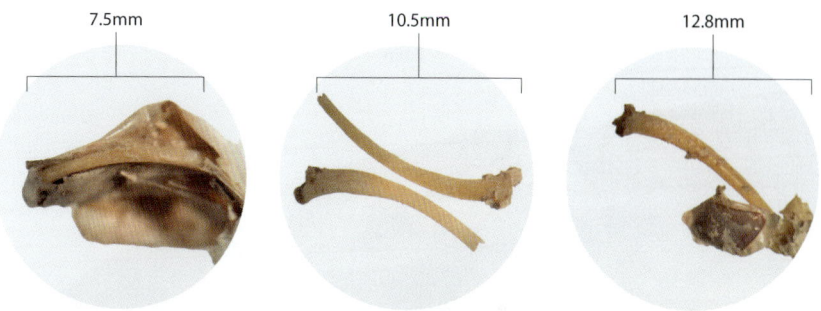

7.5mm

10.5mm

12.8mm

Ribs and scales.

4mm

A matrix of bones and skin.

Snake teeth

These images reveal the shape and arrangement of teeth in a grass snake and an adder.

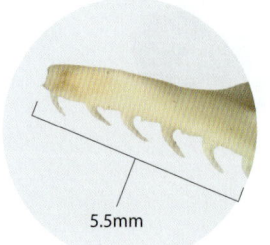

5.5mm

One of the upper left jawbones and teeth of a grass snake.

3mm

A decomposed grass snake revealing the front upper teeth. (Collected by Becky Coffin)

22mm

A decomposed grass snake head revealing the long, curved and backward-pointing teeth on one of the upper right jawbones. (Collected by Becky Coffin)

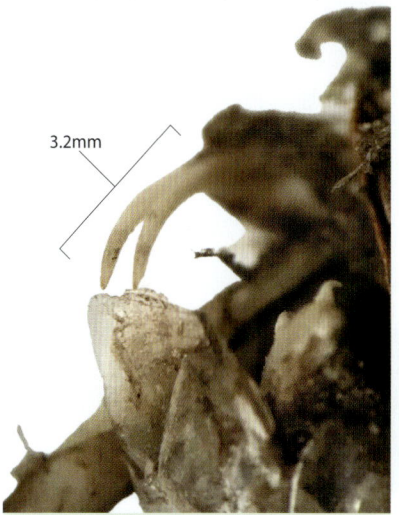

3.2mm

The fangs of a female adder.

1.4mm

The curved teeth in the jawbone of a female adder.

Common or viviparous lizard

4.8mm

The front part of the skull of a viviparous lizard revealing the pattern of similar, sharp peg-like teeth.

1mm

The lower-right jawbone of a viviparous lizard revealing the repeating peg-like teeth.

4.9mm

The front leg (ulna and radius) of a viviparous lizard.

Length of one
vertebra 1.7mm

Some vertebrae and ribs of a viviparous lizard.

The common or viviparous lizard is our most widespread and common in Britain and Ireland. It is a common prey item of birds of prey, in particular kestrels, and is also eaten by great grey and red-backed shrikes.

Pellets that contain keratinous scales, lacking bony cores, from the skin of reptiles can be distinguished from snakes if they reveal limb bones, pelvic girdles and parts of the skull. In prey studies the pelvic girdles are most robust and allow more accurate counting of individual lizards (Yalden & Warburton 1979).

Slow worm

The mummified head of a slow worm showing the sharp, backward pointing teeth. (Collected by Dr James Drewitt)

4.2mm (left side set of teeth from front to back)

The scales of a slow worm, revealing the ossified or bony cores within each one. Most of the softer non-bony scale parts have decayed (×20 magnification). (Collected by Dr James Drewitt)

Invertebrates

Although many different invertebrate species may appear in the diet of birds (and hence their pellets), there are some common species that can be easily identified and many others where this is not possible.

Some, such as those from larger insects like cockchafers, garden chafers, dung beetles, crickets, grasshoppers and dragonflies may be more easily identified. Others are beyond the scope of this book and may need to be compared with specimens in personal collections, reference collections in museums or using detailed identification keys – although the latter are often better designed for whole specimens (Losito & Cowley 2020). Some may be identified to species level, while others may only be identified to a genus, family or order. For many species, such as crickets and grasshoppers, the number of individuals in a given pellet can be estimated by counting the number of mandibles or other non-digestible pieces (Catry *et al.* 2016). The quantity of earthworms eaten can be estimated by counting and measuring their chaetae under a microscope (seen as tiny, shiny golden bristles) or by weighing pellets for their sand content, assuming 1g of sand comes from 10g of earthworms (Yalden & Warburton 1979; Yalden 2009). If you have a gritty pellet, Yalden (2009) suggests weighing each pellet before breaking it apart. If a pellet weighs 1 gram and contains 75% fur and 25% grit, then assume that 25% of the 1 gram originates from earthworms eaten. Once all pellets have been weighed and their sand content ascertained, each individual mass of sand can be added together. The total figure can be multiplied by 10 to get an estimate of the weight of earthworms eaten. This can then be compared with the estimate of other animals found in the pellets. Yalden also highlights that there are some errors, such as sand being more dense and therefore heavier than fur. However, having some quantitative data on earthworm prey, even with some error margin, is better than nothing at all. Estimating earthworms in animal diets is detailed further in Wroot (1985).

John Walters and Mark Telfer have produced some excellent guides to identifying a range of British ground beetles, which are often the most obvious parts of insects in pellets: http://johnwalters.co.uk/publications/guide-to-british-beetles.php

The following book is out of print, although may be sourced through second-hand online bookshops and through university libraries.
- Shiel, C., McAney, C., Sullivan, C. and Fairley, J. 1997. *Identification of arthropod fragments in bat droppings.* Mammal Society Occasional Publication 17.

The invertebrates below are some of the more obvious of such prey items that may appear in a pellet.

Cockchafer

A female cockchafer – smaller and more compact than a male.

A male cockchafer, larger than females, with more elaborate, fan-like antennae.

At certain times of the year pellets from birds such as tawny owls, kestrels, shrikes and corvids will contain the remains of cockchafers. These coincide with when cockchafers suddenly appear and may be locally abundant. Their chestnut-brown legs, wing cases and antennae are easily visible among other insects and hair or feathers in a pellet.

Rose chafer

A rose chafer, which are often found feeding on nectar and pollen in gardens.

Rose chafers are large, distinctive beetles with bright, iridescent green exoskeletons. Like the cockchafer, they suddenly appear in spring or summer and may be locally common and easily foraged by birds.

Woodland dor beetle (dung beetle)

A woodland dor beetle, a type of dung beetle.

The woodland dor beetle and other similar-looking species, such as the common dor beetle and common dumbledor beetle, are often found in the pellets of little owls, tawny owls and corvids, such as jackdaws. They have large, distinctive legs, mouthparts and wing cases which are often black with iridescent purple-blue colours.

Mud snails

A mud snail feeding on green seaweed. Shell length 4.2mm. (Ian Smith – Flickr profile: Morddyn)

Mud snails, also known as laver spire shells, are commonly found in many thousands per square metre on the surface of sand and mud of estuaries and beaches that are exposed at low tide. They are a common food source for many species of waders, including curlew, redshank and dunlin. Their current Latin name is *Peringia ulvae*, although many will know this snail by its previous name, *Hydrobia ulvae* (which will be prevalent in books and peer-reviewed papers).

Sea slater

A sea slater on seashore rocks above the high-tide mark. Not to scale. (Malcolm Storey, bioimages.org.uk)

Similar to a woodlouse, the much larger sea slaters can be found on the beach living under and among rocks and debris washed up by the tide. Waders, such as curlews, probe among rocks and piles of seaweed to find sea slaters and other invertebrates such as sandhoppers. They may be hunted more easily by waders at night when they come out from their crevices and hiding places.

Polychaete worms

The jaws of the free-swimming king ragworm found in shallow water among seaweeds on the ebb spring tide. Not to scale. (Malcolm Storey, bioimages.org.uk)

The jaws of marine polychaete worms, especially the free-swimming king ragworm, may appear in pellets either as secondary consumed prey or as directly consumed by birds such as cormorants, curlews and green sandpipers. When they are in abundance during their swarming season, or too large to have been swallowed by fish, it is highly likely they have been consumed by the bird producing the pellet, such as a cormorant. The jaws preserve better than fish otoliths, remaining unchanged even after being in a bird's stomach. Once extracted from pellets, they can be paired up according to their size, shape and wear. Their size can determine how large the worm was when it was eaten (see Leopold & Van Damme 2002).

Freshwater shrimps

A freshwater shrimp. Not to scale. (Malcolm Storey, bioimages.org.uk)

Freshwater shrimps are eaten by a range of species and have been particularly studied in the diet of green sandpipers in Britain. For example, when feeding on freshwater shrimps in watercress beds, green sandpipers feed for 15–20 minutes and then pause to preen and regurgitate pellets and/or poo (Holt & Warrington 1996). Their pellets are full of the exoskeletons of the shrimps including their appendages (legs), of which two pairs towards the head are known as gnathopods. These have more distinctive claws compared to the other appendages and can be measured to determine the size of the shrimps eaten (see Holt & Warrington 1996).

Cephalopod beaks

 The upper beak from a small *Loligo* squid. Actual size.

Squid, such as loliginids, may sometimes be eaten by seabirds (such as great black-backed gulls, cormorants and shags) and leave their telltale signs as hard beaks in pellets. Identification of species is not easy, so in many cases the presence of squid and number of beaks may be all that can be recorded. Although out of print, academic libraries and second-hand online shops may have copies of: Clarke, M.R. (ed.). 1986. *A Handbook for the Identification of Cephalopod Beaks*. Clarendon Press, Oxford.

Birds

Feathers

Feathers are common in pellets, depending on the species involved. Wing and tail feathers are mostly plucked out before the flesh is consumed, although they may still be present. The smaller, fluffier feathers may be both plucked and discarded or swallowed. In grebes, feathers are eaten on purpose and specifically fed to their chicks to help keep food in the stomach until it is digested. They also help retain undigestible food and parasites before forming them as a pellet. The feathers work as a feather plug-sieve. However, they do not appear to be essential; of the very few pellets seen regurgitated by grebes not all appeared to contain feathers (Simmons 1956; Storer 1961; Kop 1972; Piersma & Eerden 1989; Jehl 2017).

In pellets, the smaller feathers are more difficult to attribute to species as their structure and quality is damaged from the digestive process, although those with striking patterns, such as water rail, or vivid colours, such as greenfinch, can be more easily identified. The white-grey body feathers of pigeons often dominate the pellets of peregrines. Small feathers that are largely indistinguishable to the naked eye can be examined under the microscope.

A paper by Tim Brom outlines how to identify feathers by their microscopic barbule (hooks that bind the individual filaments together) patterns and lengths: Brom, T.G. 1986. Microscopic identification of feathers and feather fragments of Palearctic birds. *Bijdragen tot de Dierkunde* 56(2): 181–204.

Intact large feathers can be identified using a range of books and an online resource:
- Brown, R., Lees, D., Ferguson, J. and Lawrence, M. 2021. *Tracks and Signs of the Birds of Britain and Europe*. Bloomsbury Publishing, London.
- Featherbase is an online database of feathers of each bird species found in the UK, across Europe and other parts of the world, featherbase.info
- Fraigneau, C. 2021. *Feathers: An Identification Guide to the Feathers of Western European Birds*. Helm, London (translated by Tony Williams).

Bird bones (excluding skulls)

Many avian predators and scavengers eat birds whole or in parts, meaning that their bones, alongside feathers, can be discovered in pellets. The length and shape of these bones are usually distinctive and different to small mammal bones, in particular the humerus, breastbone, pelvis and leg bones. If the skull is in a pellet then identification is more likely. A range of commonly found skulls are featured in this chapter. In the absence of any skull, finding a humerus can still help assign the bird a probable size and mass (and thereby narrow down the potential species), by using the following formula (Morris & Burgis 1988; Yalden 2009):

$$ln \text{ weight (g)} = (2.4221 \times ln \text{ humerus length (mm)}) - 3.8027$$

Key:
- $ln(x) = \log_e(x)$ = the natural logarithm of x
- e^x = exponential function raised to the power of x

Which buttons to use on a scientific calculator

The buttons *ln* and e^x can be used for the calculation and are usually pressed first before inputting the relevant number.

Which buttons to use on a smart phone

On a smartphone, the scientific calculator can often be found by going into calculator and rotating the phone so the screen is in landscape mode. Depending on the smartphone you may need to press the *ln* button first before inputting a number, while on other models you may need to press this after putting in the number. Just try it out and see what happens.

For example, measure the length of a bird humerus bone in millimetres. Type this length into the calculator and press *ln* to get the *ln* of the number. Once you have done the full calculation using this number ((2.4221 × *ln* humerus length (mm)) – 3.8027) you will end up with the *ln* weight. You can put this number into the calculator and press the button e^x to get the predicted weight of the bird. Sometimes the e^x button needs to be used before the number.

To check if your calculation has been done correctly and is giving you a realistic prediction of weight see Table 2 in Morris & Burgis (1988), which outlines the body weights of birds predicted from different humerus measurements (such as 5mm, 10mm, 20mm, etc.).

If humeri bones are not present in the pellet, then the size of other bones present may allow the bird to be assigned to the category of small (sparrow-size), medium (blackbird-size) or large (pigeon-size) and thus still be used in an analysis.

Humerus

The humerus has a distinctive broad, three-dimensional pitted joint (for muscle attachment).

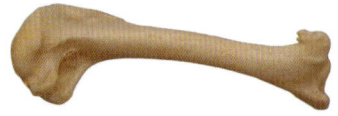

A magpie humerus.

A yellowhammer humerus. (From a barn owl pellet, University of Bristol)

Breast or keel bone

The breast or keel bone (also known as the sternum) is a distinctive three-dimensional bone where the main wing muscles attach.

Magpie breastbones.

Yellowhammer breastbone. (From a barn owl pellet, University of Bristol)

[All images are life size unless otherwise indicated.]

Pelvic girdle

The pelvic girdle of a magpie viewed from the side and from beneath. This is a long, narrow bone, bowed on each side and comprising fused vertebrae, the socket joints for the femurs and long thin tips on each side of one end. It is often confused for a skull or as being from another sort of animal.

Carpometacarpus

A carpometacarpus from a magpie. This distinctive single complex bone is a combination of fused wing bones comprising the first, second and third metacarpals. It forms part of the outer portion of the wing where the primary wing feathers grow.

Furcula

The furcula from a magpie. The furcula is a V-shaped structure made of two thin bones, the clavicles. These are the equivalent of our collarbone and are often known as the wishbone.

Coracoid

The coracoid from a magpie. This is part of the shoulder joint and has a distinctive flattened, triangle-shaped tip. It connects the humerus to the breastbone.

Leg bones

The leg bones of birds are generally elongated, especially the tibiotarsus and fibula, which are the equivalent of human shin bones (below the knee) and mostly hidden inside the body or surrounded by muscle and feathers. Connected to this is the most visible part of the leg, the tarsometatarsus, a long ankle bone that is variable in length depending on the species and covered in coloured scales or feathers. This then connects to the various toe bones. The femur, with the ball-and-socket joint, is hidden inside a bird's body, connected to the pelvic girdle and the tibiotarsus and fibula.

If a bird of prey has swallowed the legs individually or consumed a whole bird, the leg bones tend to be obvious in a pellet due to their length. For some, such as waders eaten by a barn owl, the legs may still be intact and covered in their horny, scaly skin.

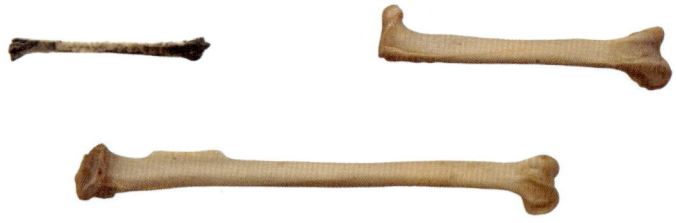

The tarsometatarsus from a yellowhammer (top left) and a femur (top right) and tibiotarsus (bottom) from a magpie.

Bird skulls

A barn owl pellet containing the skull of hoopoe, Fuerteventura, Canary Islands. (Guillermo Delgado Castro/Aurora Gonzalo Torodo, Museum of Nature and Archaeology (Museo de Naturaleza y Arqueología))

[All images are life size unless otherwise indicated.]

A barn owl pellet containing the skull of a corn bunting, Fuerteventura, Canary Islands. (Guillermo Delgado Castro/Aurora Gonzalo Torodo, Museum of Nature and Archaeology (Museo de Naturaleza y Arqueología))

The heads of birds are often discarded at or below the feeding area of the predator. However, they may also be swallowed from time to time and regurgitated in pellets, just like small mammal skulls.

The species shown below are those more likely to be found in pellets. Brown *et al.* (2021) also feature a wide selection of bird skulls and measurements, in particular proportions of bills versus overall skull length and in relation to the cranium or brain case. The origin of some skulls is not given as they are either from mixed sources of peregrine prey or unknown sources.

Wren

Skull length range: 29–32mm (Brown *et al.* 2021).

Long thin angled bill and small, delicate cranium.

Robin

Skull length range: 32–34mm (Brown *et al.* 2021).

Larger skull than wren with a slightly thicker, less elongated bill and longer nostril.

Dunnock

Skull length range: 30–34mm (Brown *et al.* 2021).

Similar in size and form to robin. Thicker, stouter bill compared to robin.

[All images are life size unless otherwise indicated.]

Blackbird

Skull length range: 47–52mm (Brown *et al.* 2021).

Compared to starling, has a thicker and straighter bill. Less tapered at the end and broader base where bill meets the cranium.

Goldcrest

Skull length range: 24–25mm (Brown *et al.* 2021).

A tiny teardrop-shaped skull with a delicate thin tweezer-like bill which is not as elongated as a wren.

Chiffchaff

Skull length range: 27–28mm (Brown *et al.* 2021).

Resembles goldcrest, although slightly longer in overall length.

Yellowhammer

Skull length range: 28–31mm (Brown *et al.* 2021).

The kinked upper bill of the yellowhammer is distinctive (and shared with reed buntings which inhabit more wetland environments). (Left skull collected by Mike Rogers)

Swallow

Skull length range: 28–30mm (Brown *et al.* 2021).

Similar to other insect-feeding birds although has a wider, flattened bill base. (Peregrine prey, collected by Nick Dixon)

[All images are life size unless otherwise indicated.]

House martin

Skull length range: 23–26mm (Brown *et al.* 2021).

Similar to swallow with a wide flattened bill base. Larger nostrils compared to swallow. (Peregrine prey, collected by Nick Dixon)

Starling

Skull length range: 49–56mm (Brown *et al.* 2021).

Commonly found in urban and rural environments, especially close to or at large winter roosts. Angled and tapering bill. (Peregrine prey, collected by Nick Dixon)

The leg of a starling, sometimes found whole in barn owl pellets. (University of Bristol)

Meadow pipit

Skull length range: 31–34mm (Brown *et al.* 2021).

A common prey item especially in upland (moorland) and coastal habitats. Long, thin bill and narrow teardrop-shaped cranium. (Peregrine prey, collected by Nick Dixon)

[All images are life size unless otherwise indicated.]

Skylark

Skull length range: 35–37mm (Brown *et al.* 2021).

A common prey item, especially in open country, upland (moorland) and arable environments. Larger skull and thicker, stouter bill compared with meadow pipit.

Greenfinch

Skull length range: 31–32mm (Brown *et al.* 2021).

Large, thick and robust bill compared to other finches featured.

As in this photo, the bill is often removed by a predator such as a peregrine. (Collected by Nick Dixon)

Chaffinch

Skull length range: 26–30mm (Brown *et al.* 2021).

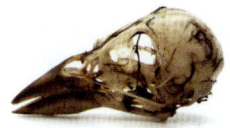

Bill less thick and more narrow compared to greenfinch.

Goldfinch

Skull length range: 28–30mm (Brown *et al.* 2021).

Narrow, tapering bill, ending in finer point compared to other finches. (Collected by Denise Wawman)

[All images are life size unless otherwise indicated.]

Siskin

Skull length range: 23–27mm (Brown *et al.* 2021).

Skull and bill smaller than goldfinch and shorter, less elongated bill. (Collected by Denise Wawman)

Linnet

Skull length range: 24–28mm (Brown *et al.* 2021).

Skull similar size to goldfinch with a shorter and thicker bill.

Bullfinch

Skull length range: 27–28mm (Brown *et al.* 2021).

Rounded cranium and bill thick, short and compact.

House sparrow

Skull length range: 29–31mm (Brown *et al.* 2021).

Thick robust bill and easily confused with greenfinch, although slightly smaller and lighter in build. (Collected by Mike Dilger)

Great tit

Skull length range: 28–32mm (Brown *et al.* 2021).

Stout bill with a smooth, round cranium. Larger skull and longer bill compared with blue tit. The two skulls above show the difference between one with a bill sheath (right) and the other missing this element (left).

[All images are life size unless otherwise indicated.]

Blue tit

Skull length range: 24–25mm (Brown *et al.* 2021).

Short, stout bill with a smooth, round cranium.

Feral pigeon

The skull length of feral pigeons found living in urban, suburban and industrial areas are in the range 51–54mm. More extreme varieties kept as loft or racing pigeons may have longer or shorter skulls (30mm or even less).

Intermediate in size between woodpigeon and collared dove. (Peregrine prey, collected by Nick Dixon)

Woodpigeon

Skull length range: 55–60mm (Brown *et al.* 2021).

Large pigeon skull with thick bill.

Collared dove

Skull length range: 46–49mm (Brown *et al.* 2021).

Smaller and thinner bill compared to other common pigeons and doves. (Peregrine prey, collected by Nick Dixon)

Stock dove

Skull length range: 51–54mm (Brown *et al.* 2021).

Similar in size to feral pigeon and easily misidentified if lacking the yellow bill sheath. (Peregrine prey)

Kingfisher

Skull length range: 52–69mm (Brown *et al.* 2021).

Elongated pointed bill and flattened cranium.

[All images are life size unless otherwise indicated.]

Little grebe
Skull length range: 45–53mm (Brown *et al.* 2021).

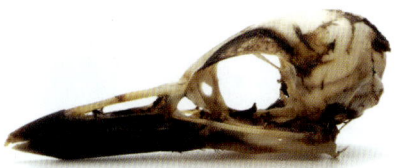

Stout bill with outer portion of lower bill angled. (Peregrine prey)

Snipe
Skull length range: 94–95mm (Brown *et al.* 2021).

Very long, thin bill and short, rounded cranium (often damaged). (Peregrine prey, collected by Nick Dixon)

Water rail
Skull length range: 67–72mm (Brown *et al.* 2021).

Thin, elongated bill that curves downwards and tapers to a blunt tip. The bill sheath is likely to be missing in a pellet and may vary in colour. While bright red in adults, the bill sheath colour easily fades and is more likely to be brown or yellow-brown.

[All images are life size unless otherwise indicated.]

Storm petrel

Skull length range: 31–34mm.

Resembles that of a small dove. Storm petrels have a rounded cranium and pigeon- or lapwing-like bill (when the black sheath is absent). This species comes to shore along rocky islands and coastlines at dusk and during the night, where they may sometimes be targeted by predators such as barn owls, gulls and skuas. Molène archipelago, Brittany, France. (From barn owl pellets, Bernard Cadiou)

Leach's petrel

Skull length range: 40–43mm (Post 1998).

Resembles storm petrel although larger. May be targeted by skuas and gulls as it returns to its nesting burrow at night.

Fish

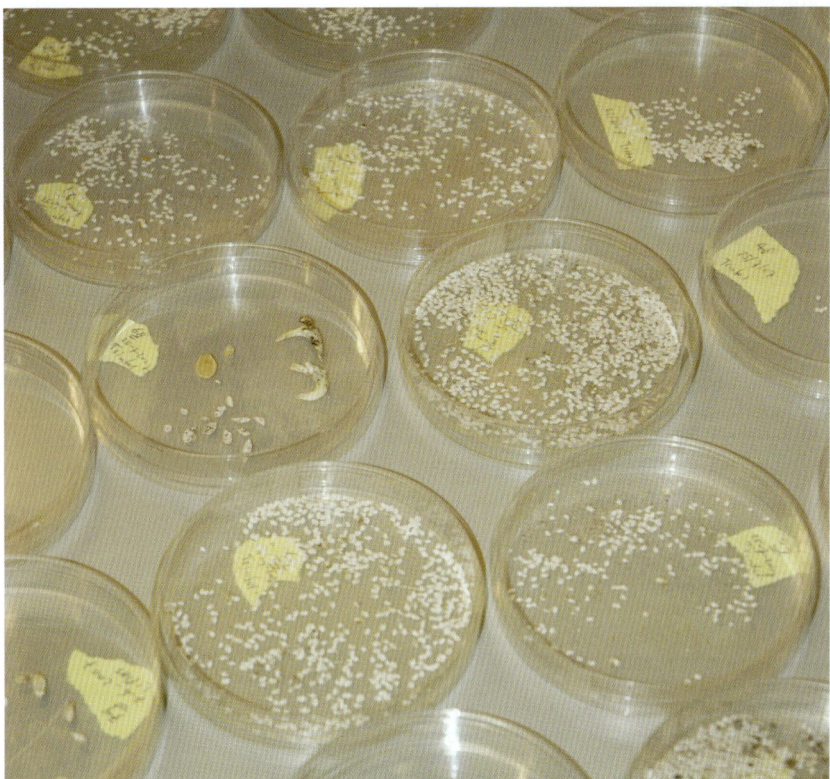

Fish otoliths – mostly from ruffe – extracted from cormorant pellets from the Lake IJsselmeer area in the Netherlands. (Stef van Rijn)

The pellets of kingfishers mostly comprise fishbones such as vertebrae, pharyngeal bones, jawbones and scales. These can be extracted and identified using reference collections and published keys/guides (see Reynolds & Hinge 1996, Čech & Čech 2015 and Nessi 2021).

Seabird pellets from cormorants, shags, terns and gulls often contain multiple bones of fish. The remains of jawbones (containing teeth), vertebrae and ear bones, known as otoliths, from pellets can be used to find out what the birds have been feeding on.

Otoliths survive the digestion process and often appear in seabird pellets and poo. They have intricate shapes and patterns that are unique to fish species, while their size relates to the age of the fish. The acidic digestive juices of birds such as cormorants may reduce their size and need to be taken into account when analysing these remains to estimate the size of fish eaten (Martucci *et al.* 1993).

Exploring the bones of fish is a labour-intensive and detailed process requiring expertise and the use of fish ear bone databases and atlases, reference collections and other published information.

Below are a selection of websites and resources that can help with this process of identification:

- Centre for Environment Fisheries and Aquaculture Science offer services that identify and age otoliths as well as offering training in how to do this: cefas.co.uk/services/laboratory-services-and-analysis/age-determination -and-otolith-science
- Leopold M.F., van Damme, C.J.G., Philippart, C.J.M. and Winter, C.J.N. 2001. *Otoliths of North Sea Fish 1.0. An Interactive guide of identification of fish from the SE North Sea, Wadden Sea and adjacent fresh waters by means of otoliths and other hard parts*. https://otoliths-northsea.linnaeus.naturalis.nl/linnaeus_ng/app/views/introduction/topic.php?id=3327&epi=87
- Available through nhbs.com: Härkönen, T. 1986. *Guide to the Otoliths of the Bony Fishes of the Northeast Atlantic*. Danbiu ApS, Hellernp, Denmark.
- Although out of print, the following may be available second-hand through websites and marine research libraries: Campana, S.E. 2004. *Photographic Atlas of Fish Otoliths of the Northwest Atlantic Ocean* (No. 133). NRC Research press.
- Alonso, H., Granadeiro, J.P., Ramos, J.A. and Catry, P. 2013. Use the backbone of your samples: fish vertebrae reduces biases associated with otoliths in seabird diet studies. *Journal of Ornithology* 154: 883–886.
- For contents found in herring and lesser black-backed gulls (and other seabirds) feeding on marine animals: Hunt, V. 2006. *Larus fuscus* and *Larus argentatus*: pellet and bolus analysis guide. Unpublished report, Royal Netherlands Institute for Sea Research, Texel.

13 WHAT ELSE MIGHT YOU FIND IN A PELLET?

When dissecting pellets you may come across things that were not originally in the pellet, in particular animals that like to eat the hair and feathers found in them. You may encounter the larvae of moths that specialise in eating what is left in a pellet (see more below). They look like white grubs, maggots or caterpillars with an orange-brown head. They are also found in old nests of birds and provide food for birds that specialise on feeding in and around hollows (and nest in them). These animals are an important part of the food chain, consuming the dried remains of animals that few others will eat and themselves providing food for birds. They are known as keratophagous insects.

How can I stop my pellets from being eaten by moths and beetles?

Freezing pellets for at least a month will kill any small animals feeding on the pellets. It is worth freezing several times to ensure eggs or larvae do not survive (as freezing just once can mimic a cold winter and still see some survive and emerge). If you are keeping some pellets whole for reference, freezing them first is a good way of ensuring they do not get eaten and disintegrate over time. They will need to be kept dry in sealed bags or containers once frozen and allowed to return to room temperature. Ensure they are dried out before putting in a container, otherwise they will sweat and go mouldy. It is well worth labelling any bags with the location and date they were found or collected.

What are the clues that pellets are being eaten?

As you start to poke around and open up pellets it will soon become obvious if there are creatures inside. They are harmless to us and can just be put to one side and then released outside. If the pellets have been frozen they will most likely be dead anyway (although it is worth checking and re-freezing if there are newly emerged larvae).

The common larvae found in bird pellets are featured below. For more information about them and further photos look up the 'Clothes moths identification guide' on London's Natural History Museum website.

Caterpillars or larvae of micromoths

There are several species of micromoth that specialise in feeding on pellets and bird nests (and anything that is made of keratin); these include the common clothes moth and the case-bearing clothes moth. These insects do not need any water and feed on the dry hair, feather and other similar parts of an animal (as well as items in the home that are made from natural fibres such as wool and cotton). The larvae are creamy-white with an orange-brown head and up to 13mm long. The adults are tiny (wingspan 12–17mm) with powdery grey-brown colours. A collection of cocoons may protrude out of the pellets in which they were developing into moths or have already emerged as such.

Carpet beetles

The hairy larvae of carpet beetles, known as 'woolly bears', may also be found in pellets. The varied carpet beetle is one of the most commonly encountered species. Its small, compact furry larvae are 4–4.5mm long and will eat what is left in a pellet (and often destroy museum collections).

(Pascal – Flickr profile: pasukaru76, taken from Flickr. CC0 1.0 Universal)

The poo or frass of larvae

The frass or poo of insect larvae feeding on pellets often look like fine, powdery poppy seeds. They are hard, often rounded and left among the disintegrating pellets as they feed.

The frass (tiny dark spheres) from insect larvae that have fed on the freeze-dried remains of a common toad. (Bristol Museum & Art Gallery)

14 A FINAL REFLECTION

I have strived for this book to provide all the detail you need to identify pellets that you may find and/or the bony remains often found inside them, particularly the skulls and other bones of small mammals and birds. I hope that it helps support you in your wildlife discoveries, whether that be at home, at an event or through more detailed research work. There are certainly some fascinating things to find in pellets!

I would like to thank everyone who has provided information, photos and pellets for this book. Individual credits are included in the photo captions and any omissions, of which I hope there are none, are purely accidental. I am incredibly grateful to you all and hope that you also enjoy and find this book useful.

REFERENCES

Acampora, H., Berrow, S., Newton, S. and O'Connor, I. 2017. Presence of plastic litter in pellets from Great Cormorant (*Phalacrocorax carbo*) in Ireland. *Marine Pollution Bulletin* 117(1–2): 512–514. https://doi.org/10.1016/j.marpolbul.2017.02.015

Anděra, M. and Horacek, I. 1986. Dormice (Gliridae) in Czechoslovakia. Part I: *Glis glis, Eliomys quercinus* (Rodentia: Mammalia). Folia Musei Rerum Naturalium Bohemiae Occidentalis, Plzeň, *Zoologica* 24: 3–47.

Andersson, M. and Götmark, F. 1980. Social organization and foraging ecology in the arctic skua *Stercorarius parasiticus*: a test of the food defendability hypothesis. *Oikos* 35(1): 63–71. https://doi.org/10.2307/3544727

Andreotti, A. and Borghesi, F. 2013. Embedded lead shot in European starlings *Sturnus vulgaris*: an underestimated hazard for humans and birds of prey. *European Journal of Wildlife Research* 59(5): 705–712.

Barlow, K.E., Jones, G. and Barratt, E.M. 1997. Can skull morphology be used to predict ecological relationships between bat species? A test using two cryptic species of pipistrelle. *Proceedings of the Royal Society of London Series B* 264(1388): 1695–1700. https://doi.org/10.1098/rspb.1997.0235

Barn Owl Trust. 2012. *Barn Owl Conservation Handbook*. Pelagic Publishing, Exeter.

Barton, N.W. and Houston, D.C. 1993. A comparison of digestive efficiency in birds of prey. *Ibis* 135(4): 363–371. https://doi.org/10.1111/j.1474-919X.1993.tb02107.x

Barrett, R.T., Camphuysen, K., Anker-Nilssen, T., Chardine, J.W., Hairness, R.W., Garthe, S., Hüppop, O., Leopold, M.F., Montevecchi, W.A. and Veit, R.R. 2007. Diet studies of seabirds: a review and recommendations. *ICES Journal of Marine Science* 64(9): 1675–1691. https://doi.org/10.1093/icesjms/fsm152

Below, T.H. 1979. First reports of pellet ejection in 11 species. *The Wilson Bulletin* 91(4): 626–628.

Bijlsma, R.G. 1999. Do honey-buzzards *Pernis apivorus* produce pellets? *Limosa* 72: 99–103.

Birkhead, T. 1991. *The Magpies*. T. & A.D. Poyser, London.

Birkhead, T.R. and Clarkson, K. 1985. The magpie as an aid to teaching behaviour and ecology. *Journal of Biological Education* 19(2): 163–168. https://doi.org/10.1080/00219266.1985.9654715

Booms, T.L. and Fuller, M.R. 2003. Gyrfalcon diet in central west Greenland during the nesting period. *Condor* 105(3): 528–537. https://doi.org/10.1093/condor/105.3.528

Bresgunova, O.A. 2014. О ночёвках соек (*Garrulus glandarius*) в г. Харькове [Notes on Jay (*Garrulus glandarius*) roosts in Kharkov city]. *The Birds of Siverskyi Donets River Basin* 12: 284–291.

Brown, J.C. and Twigg, G.I. 1969. Studies on the pelvis in British Muridae and Cricetidae (Rodentia). *Journal of Zoology* 158: 286–292. https://doi.org/10.1111/j.1469-7998.1969.tb04967.x

Brown, A., Gilbert, G. and Wotton, S. 2012. Bitterns and bittern conservation in the UK. *British Birds* 105(2): 58–87.

Brown, R.W., Lawrence, M.J. and Pope, J. 2012. *Animals Tracks, Trails & Signs*. Octopus Publishing Group Ltd, London.

Brown, R., Lees, D., Ferguson, J. and Lawrence, M. 2021. *Tracks and Signs of the Birds of Britain and Europe*. Bloomsbury Publishing, London.

Buatip, S., Karntanut, W. and Swennen, C. 2014. Food Habits of Little Egrets (*Egretta garzetta*) at a Colony in Pattani, Southern Thailand. *Natural History Bulletin of the Siam Society* 60(1).

Buckley, N.J. 1990. Diet and feeding ecology of great black-backed gulls (*Larus marinus*) at a southern Irish breeding colony. *Journal of Zoology* 222(3): 363–373. https://doi. org/10.1111/j.1469-7998.1969.tb04967.x

Bugoni, L. and Vooren, C.M. 2004. Feeding ecology of the Common Tern *Sterna hirundo* in a wintering area in southern Brazil. *Ibis* 146(3): 438–453.

Catry, T., Ramos, J.A., Paiva, V.H., Martins, J., Almeida, A., Palma, J., Andrade, P.J., Peste, F., Trigo, S. and Luís, A. 2006. Intercolony and annual differences in the diet and feeding ecology of little tern adults and chicks in Portugal. *Condor* 108(2): 366–376. https://doi. org/10.1093/condor/108.2.366

Catry, I., Catry, T., Alho, M., Franco, A.M.A. and Moreira, F. 2016. Sexual and parent–offspring dietary segregation in a colonial raptor as revealed by stable isotopes. *Journal of Zoology* 299(1): 58–67. https://doi.org/10.1111/jzo.12324

Čech, M. and Čech, P., 2015. Non-fish prey in the diet of an exclusive fish-eater: the Common Kingfisher *Alcedo atthis*. *Bird Study* 62(4): 457–465. https://doi.org/10.1080/00063657. 2015.1073679

Chas, A.U. and Carter, T.D. 1925. Notes on Two Ground-Nesting Birds of Prey. *The Auk* 42(1): 31–41. https://doi.org/10.2307/4074304

Clarke, R., Bourgonje, A. and Castelijns, H. 1993. Food niches of sympatric Marsh Harriers *Circus aeruginosas* and Hen Harriers *C. cyaneus* on the Dutch coast in winter. *Ibis* 135(4): 424–431. https://doi.org/10.1111/j.1474-919X.1993.tb02115.x

Colling, A.W. and Brown, E.B. 1946. The breeding of Marsh and Montagu's Harriers in North Wales in 1945. *British Birds* 39: 233–243.

D'Souza, J.M., Windsor, F.M., Santillo, D. and Ormerod, S.J. 2020. Food web transfer of plastics to an apex riverine predator. *Global Change Biology* 26(7): 3846–3857. https:// doi.org/10.1111/gcb.15139

Dodson, P. and Wexlar, D. 1979. Taphonomic investigations of owl pellets. *Paleobiology* 5(3): 275–284. https://doi.org/10.1017/S0094837300006564

Duke, G.E. 1977. Pellet egestion by a captive Chimney Swift (*Chaetura pelagica*). *The Auk* 94(2): 385–385.

Figueroa, R.A. and Stappung, E.S.C. 2003. Food of breeding Great White Egrets in an agricultural area of southern Chile. *Waterbirds* 26(3): 370–375. https://doi.org/10.1675/ 1524-4695(2003)026[0370:FOBGWE]2.0.CO;2

Francour, P. and Thibault, J.C. 1996. The diet of breeding osprey *Pandion haliaetus* on Corsica: exploitation of a coastal marine environment. *Bird Study* 43(2): 129–133. https:// doi.org/10.1080/00006359609461004

Furness, R.W. 1987. *The Skuas*. Poyser Press, Calton.

Gagliardi, A., Martinoli, A., Wauters, L. and Tosi, G. 2003. A floating platform: a solution to collecting pellets when cormorants roost over water. *Waterbirds* 26: 54–55. https://doi. org/10.1675/1524-4695(2003)026[0054:AFPAST]2.0.CO;2

Giles, N. 1981. Summer diet of the grey heron. *Scottish Birds* 11: 153–159.

Gill, B.J. 1980. Foods of the shining cuckoo (*Chrysococcyx lucidus*, Aves: Cuculidae) in New Zealand. *New Zealand Journal of Ecology* 3: 138–140.

Glue, D.E. 1977. Feeding ecology of the Short-eared Owl in Britain and Ireland. *Bird Study* 24(2): 70–78. https://doi.org/10.1080/00063657709476536

Goss-Custard, J.D. and Jones, R.E. 1976. The diets of redshank and curlew. *Bird Study* 23(3): 233–243. https://doi.org/10.1080/00063657609476507

Granadeiro, J.P., Monteiro, L.R., Silva, M.C. and Furness, R.W. 2002. Diet of common terns in the Azores, Northeast Atlantic. *Waterbirds* 25(2): 149–155. https://doi.org/10.1675/ 1524-4695(2002)025[0149:DOCTIT]2.0.CO;2

Green, A.J., Lovas-Kiss, Á., Reynolds, C., Sebastián-González, E., Silva, G.G., van Leeuwen, C.H. and Wilkinson, D.M. 2023. Dispersal of aquatic and terrestrial organisms by waterbirds: A review of current knowledge and future priorities. *Freshwater Biology* 68(2): 173–190.

Grimm, R.J. and Whitehouse, W.M. 1963. Pellet formation in a Great Horned Owl: a rodentgenographic study. *The Auk* 80(3): 301–306. https://doi.org/10.2307/4082889

Guimaraes, S., Fernandez-Jalvo, Y., Stoetzel, E., Gorgé, O., Bennett, E.A., Denys, C., Grange, T. and Geigl, E.M. 2016. Owl pellets: a wise DNA source for small mammal genetics. *Journal of Zoology* 298(1): 64–74. https://doi.org/10.1111/jzo.12285

Hammer, S., Nager, R.G., Johnson, P.C.D., Furness, R.W. and Provencher, J.F. 2016. Plastic debris in great skua (*Stercorarius skua*) pellets corresponds to seabird prey species. *Marine Pollution Bulletin* 103(1–2): 206–210. https://doi.org/10.1016/j.marpolbul.2015.12.018

Harris, S. and Yalden, D.W. 2008. *Mammals of the British Isles. Handbook 4th Edition.* The Mammal Society, Southampton.

Hacker, C.E., Hoenig, B.D., Wu, L., Cong, W., Yu, J., Dai, Y., Li, Y., Li, J., Xue, Y., Zhang, Y. and Ji, Y. 2021. Use of DNA metabarcoding of bird pellets in understanding raptor diet on the Qinghai-Tibetan Plateau of China. *Avian Research* 12(1): 1–11. https://doi.org/10.1186/s40657-021-00276-3

Häkkinen, I. 1978. Diet of the Osprey *Pandion haliaetus* in Finland. *Ornis Scandinavica* 9(1), 111–116. https://doi.org/10.2307/3676145

Harris, S. 2022. Invasions, plagues and conservation – the history of Ship Rats in Britain and Ireland. *British Wildlife* 34(3): 157–167.

Hays, H. and Cormons, G. 1974. Plastic particles found in tern pellets, on coastal beaches and at factory sites. *Marine Pollution Bulletin* 5(3): 44–46. https://doi.org/10.1016/0025-326X(74)90234-3

Holt, P. and Warrington, S. 1996. The analysis of faeces and regurgitated pellets for determining prey size: problems and bias illustrated for Green Sandpipers *Tringa ochropus* feeding on *Gammarus*. *Wader Study Group Bulletin* 79: 65–68.

Hutterer, R. 2005. Homology of unicuspids and tooth nomenclature in shrews. Advances in the Biology of Shrews, pp. 397–404. In: *Advances in the Biology of Shrews II.* Merritt, J.F., Churchfield, S., Hutterer, R. and Sheftel, B.I. (eds). Special Publication of the International Society of Shrew Biologists 1, New York.

Hardey, J., Crick, H., Wernham, C., Riley, H., Etheridge, B. and Thompson, D. 2013. *Raptors: A field guide for surveys and monitoring* (3rd edition). The Stationary Office, Edinburgh.

Hockett, B. 2018. Taphonomic comparison of bones in mountain lion scats, coyote scats, golden eagle pellets, and great-horned owl pellets. *Quaternary International* 466(B): 141–144. https://doi.org/10.1016/j.quaint.2016.02.033

Holt, D.W., Lyon, L.J. and Hale, R. 1987. Techniques for differentiating pellets of Short-eared Owls and Northern Harriers. *Condor* 89(4): 929–931. https://doi.org/10.2307/1368548

Imber, M.J. 1973. The food of grey-faced petrels (*Pterodroma macroptera gouldi* (Hutton)), with special reference to diurnal vertical migration of their prey. *The Journal of Animal Ecology* 645–662.

Jehl, J.R. 2017. Feather-eating in grebes: A 500-year conundrum. *The Wilson Journal of Ornithology* 129(3): 446–458. https://doi.org/10.1676/16-196.1

Kendall, R.J., Lacker Jr, T.E., Bunck, C., Daniel, B., Driver, C., Grue, C.E., Leighton, F., Stansley, W., Watanabe, P.G. and Whitworth, M. 1996. An ecological risk assessment of lead shot exposure in non-waterfowl avian species: upland game birds and raptors. *Environmental Toxicology and Chemistry: An International Journal* 15(1): 4–20.

Kleyheeg, E. and van Leeuwen, C.H. 2015. Regurgitation by waterfowl: an overlooked mechanism for long-distance dispersal of wetland plant seeds. *Aquatic Botany* 127: 1–5. https://doi.org/10.1016/j.aquabot.2015.06.009

Kop, P.P.A.M. 1972. Pellet-ejection by hand-reared Great Crested Grebes. *British Birds* 65: 319–321.

Kusmer, K.D. 1990. Taphonomy of owl pellet deposition. *Journal of Paleontology* 64(4): 629–637. https://doi.org/10.1017/S0022336000042669

Laudet, F., Denys, C. and Senegas, F. 2002. Owls, multirejection and completeness of prey remains: implications for small mammal taphonomy. *Acta Zoologica Cracocoviensia* 45: 341–355.

Leal, G.R., Furness, R.W., McGill, R.A., Santos, R.A. and Bugoni, L. 2017. Feeding and foraging ecology of Trindade petrels *Pterodroma arminjoniana* during the breeding period in the South Atlantic Ocean. Marine Biology 164: 1–17.

Leopold, M.F. and Van Damme, C.J. 2003. Great cormorants *Phalacrocorax carbo* and polychaetes: can worms sometimes be a major prey of a piscivorous seabird? *Marine Ornithology* 31: 83–87.

Li, L.X., Yi, X.F., Li, M.C. and Zhang, X.A. 2004. Analysis of diets of upland buzzards using stable carbon and nitrogen isotopes. *Israel Journal of Ecology and Evolution* 50(1): 75–85. https://doi.org/10.1560/MH0X-VNBG-9E4Y-KHJT

Lockie, J.D. 1956. The food and feeding behaviour of the jackdaw, rook and carrion crow. *The Journal of Animal Ecology* 25(2): 421–428. https://doi.org/10.2307/1935

Lopez, J.M. 2020. Actualistic taphonomy of barn owl pellet-derived small mammal bone accumulations in arid environments of South America. *Journal of Quaternary Science* 35(8): 1057–1069. https://doi.org/10.1002/jqs.3251

Losito, L. and Cowley, B. 2020. A survey of dung beetles (and other Coleoptera) on Lundy and an investigation of the analysis of bird pellets as a beetle survey technique. *Journal of the Lundy Field Society* 7: 39–52.

Lovas-Kiss, Á., Sánchez, M.I., Wilkinson, D.M., Coughlan, N.E., Alves, J.A. and Green, A.J. 2019. Shorebirds as important vectors for plant dispersal in Europe. *Ecography* 42(5): 956–967. https://doi.org/10.1111/ecog.04065

MacDonald, D.W. and Barrett, P. 1993. *Collins Field Guide to Mammals of Britain & Europe.* Harper Collins Publishers, London.

Marti, C.D. 1973. Food consumption and pellet formation rates in four owl species. *The Wilson Bulletin* 85(2): 178–181.

Martin, B.P. 1992. *Birds of Prey of the British Isles.* David & Charles, Newton Abbot, Devon.

Martucci, O., Pietrelli, L. and Consiglio, C. 1993. Fish otoliths as indicators of the cormorant *Phalacrocorax carbo* diet (Aves, Pelecaniformes). *Italian Journal of Zoology* 60(4): 393–396. https://doi.org/10.1080/11250009309355845

Mateo, R., Estrada, J., Paquet, J.Y., Riera, X., Domínguez, L., Guitart, R. and Martínez-Vilalta, A. 1999. Lead shot ingestion by marsh harriers *Circus aeruginosus* from the Ebro delta, Spain. *Environmental Pollution* 104(3): 435–440.

Mateo, R., Cadenas, R., Máñez, M. and Guitart, R. 2001. Lead shot ingestion in two raptor species from Doñana, Spain. *Ecotoxicology and Environmental Safety* 48(1): 6–10.

Mathews, F., Coomber, F., Wright, J. and Kendall, T. (eds). 2018. *Britain's Mammals, 2018: The Mammal Society's guide to their population and conservation status.* The Mammal Society, London.

Massa, Bruno and Rizzo, M.C. 2002. Nesting and feeding habits of the European Bee-eater (*Merops apiaster* L.) in a colony next to a beekeeping site. *Avocetta-Parma* 26(1): 25–32.

Matos, M., Alves, M., Pereira, M.R., Torres, I., Marques, S. and Fonseca, C. 2015. Clear as daylight: analysis of diurnal raptor pellets for small mammal studies. *Animal Biodiversity and Conservation* 38(1): 37–48. https://doi.org/10.32800/abc.2015.38.0037

Mauco, L., Favero, M. and Bó, M.S. 2001. Food and feeding biology of the Common Tern during the nonbreeding season in Samborombon Bay, Buenos Aires, Argentina. *Waterbirds* 24(1): 89–96. https://doi.org/10.2307/1522247

McKay, C.R. 1996. Conservation and ecology of the red-billed chough *Pyrrhocorax pyrrhocorax* (Doctoral dissertation, University of Glasgow).

Mikkola, H. 1983. *Owls of Europe*. T. & A.D. Poyser, London.

Mikula, P., Morelli, F., Lučan, R.K., Jones, D.N. and Tryjanowski, P. 2016. Bats as prey of diurnal birds: a global perspective. *Mammal Review* 46(3): 160–174. https://doi.org/10.1111/mam.12060

Morris, P.A. and Burgis, M.J. 1988. A method for estimating total body weight of avian prey items in the diet of owls. *Bird Study* 35(2): 147–152. https://doi.org/10.1080/00063658809480393

Navarro-Ramos, M.J., Green, A.J., Lovas-Kiss, A., Roman, J., Brides, K. and van Leeuwen, C.H. 2022. A predatory waterbird as a vector of plant seeds and aquatic invertebrates. *Freshwater Biology* 67(4): 657–671. https://doi.org/10.1111/fwb.13870

Nessi, A., Balestrieri, A., Winkler, A., Casoni, A.G. and Tremolada, P. 2021. Kingfisher (*Alcedo atthis*) diet and prey selection as assessed by the analysis of pellets collected under resting sites (River Ticino, north Italy). *Aquatic Ecology* 55(1): 135–147. https://doi.org/10.1007/s10452-020-09817-2

Nessi, A., Winkler, A., Tremolada, P., Saliu, F., Lasagni, M., Ghezzi, L.L.M. and Balestrieri, A. 2022. Microplastic contamination in terrestrial ecosystems: A study using barn owl (*Tyto alba*) pellets. *Chemosphere* 308(1): 36281. https://doi.org/10.1016/j.chemosphere.2022.136281

Nikolov, B.P., Kodzhabashev, N.D. and Popov, V.V. 2004. Diet composition and spatial patterns of food caching in wintering Great Grey Shrikes (*Lanius excubitor*) in Bulgaria. *Biological Letters* 41(2): 119–133.

O'Donnell, C.F. 1982. Food and feeding behaviour of the Southern Crested Grebe on the Ashburton Lakes. *Notornis* 29: 151–156.

Obuch, J. 2001. Dormice in the diet of owls in the Middle East. *Trakya University Journal of Scientific Research Series B* 2(2): 145–150.

Olsen, L. 2013. *Tracks and Signs of the Animals and Birds of Britain and Europe*. Princeton University Press, Oxfordshire. https://doi.org/10.1515/9781400847921

Piersma, T. and Eerden, M.R.V. 1989. Feather eating in Great Crested Grebes *Podiceps cristatus*: a unique solution to the problems of debris and gastric parasites in fish-eating birds. *Ibis* 131(4): 477–486. https://doi.org/10.1111/j.1474-919X.1989.tb04784.x

Post, J.N.J 1998. Biometrics of 35 specimens of the Leach's storm-petrel *Oceanodroma leucorhoa* from a wreck in southern Portugal. *Deinsea* 4(1): 77–90.

Poulin, B., Lefebvre, G. and Crivelli, A.J. 2007. The invasive red swamp crayfish as a predictor of Eurasian bittern density in the Camargue, France. *Journal of Zoology* 273(1): 98–105. https://doi.org/10.1111/j.1469-7998.2007.00304.x

Potapov, E. 2011. Gyrfalcon diet: spatial and temporal variation. In: *Gyrfalcons and Ptarmigan in a Changing World*. R.T. Watson, R.T., Cade, T.J., Fuller, M., Hunt, G. and Potapov, E. (eds). The Peregrine Fund, Boise, Idaho, USA. 1: 55–64. https://doi.org/10.4080/gpcw.2011.0106

Pretelli, M.G., Josens, M.L. and Escalante, A.H. 2012. Breeding biology at a mixed-species colony of great egret and cocoi heron in a pampas wetland of Argentina. *Waterbirds* 35(1): 35–43. https://doi.org/10.1675/063.035.0104

Ratcliffe, D.A. 1997. *The Raven*. T. & A.D. Poyser, London.

Resano-Mayor, J., Hernández-Matías, A., Real, J., Parés, F., Inger, R. and Bearhop, S. 2014. Comparing pellet and stable isotope analyses of nestling Bonelli's Eagle *Aquila fasciata* diet. *Ibis* 156(1): 176–188. https://doi.org/10.1111/ibi.12095

Reynolds, S.J. and Hinge, M.D.C. 1996. Foods brought to the nest by breeding Kingfishers *Alcedo atthis* in the New Forest of southern England. *Bird Study* 43(1): 96–102. https://doi.org/10.1080/00063659609460999

Robinson, S.A., Forbes, M.R. and Hebert, C.E. 2008. Is the ingestion of small stones by double-crested cormorants a self-medication behavior. *Condor* 110(4): 782–785. https://doi.org/10.1525/cond.2008.8541

Rolando, A. 1998. Factors affecting movements and home ranges in the jay (*Garrulus glandarius*). *Journal of Zoology* 246(3): 249–257. https://doi.org/10.1111/j.1469-7998.1998.tb00155.x

Ronayne, S.T. and Sleeman, D.P. 2013. Analysis of Barn Owl Tyto alba pellets, with notes on the sexing of Pygmy Shrew *Sorex minutus* prey in County Cork. *Irish Birds* 9: 635–636.

Salazar, R.D., Riddiford, N.J. and Vicens, P. 2005. Estudi comparatiu de la dieta de l'esplug-abous (*Bubulcus ibis*) i l'agró blanc (*Egretta garzetta*) al Parc. [A comparative dietary study of Cattle Egrets (*Bubulcus ibis*) and Little Egrets (*Egretta garzetta*) in S'Albufera Natural Park, Mallorca]. *Bolletí de la Societat d'Història Natural de les Balears*: 153–162.

Shimizu, T., Natsukawa, H., Yuasa, H., Kuroda, H. and Ichinose, T. 2022. DNA Metabarcoding Analysis of Long-Eared Owl *Asio otus* Pellets Reveals Small Animals as Its Prey. *Keio SFC Journal* 22(2): 336–347.

Simmons, K.E.L. 1956. Feather-eating and pellet-formation in the Great Crested Grebe. *British Birds* 49: 432–435.

Smith, C.R. and Richmond, M.E. 1972. Factors influencing pellet egestion and gastric pH in the barn owl. *The Wilson Bulletin*: 179–186.

Smith, C. and Jones, T. 2006. *Breeding ecology and diet of great and Arctic skuas on Handa Island 2006*. Unpublished report to SWT, SNH, The Seabird Group, JNCC.

Soler, J.J. and Soler, M. 1993. Diet of the Red-billed Chough *Pyrrhocorax pyrrhocorax* in south-east Spain. *Bird Study* 40(3): 216–222. https://doi.org/10.1080/00063659309477186

Speakman, J.R. 1991. The impact of predation by birds on bat populations in the British Isles. *Mammal Review* 21(3): 123–142. https://doi.org/10.1111/j.1365-2907.1991.tb00114.x

Stuart, C. and Stuart, M. 2013. A Field Guide to the Tracks & Signs of Southern, Central & East African Wildlife. Struik Nature, South Africa.

Storer, R.W. 1961. Observations of pellet-casting by horned and pied-billed grebes. *The Auk* 78(1): 90–92. https://doi.org/10.2307/4082237

Tarshis, I.B. and Rattner, B.A. 1982. Accumulation of 14 C-naphthalene in the tissues of redhead ducks fed oil-contaminated crayfish. *Archives of Environmental Contamination and Toxicology* 11: 155–159. https://doi.org/10.1007/BF01054891

Tatner, P. 1983. The diet of urban magpies *Pica pica*. *Ibis* 125(1): 90–107. https://doi.org/10.1111/j.1474-919X.1983.tb03086.x

Terry, R.C. 2004. Owl pellet taphonomy: a preliminary study of the post-regurgitation taphonomic history of pellets in a temperate forest. *Palaios* 19(5): 497–506. https://doi.org/10.1669/0883-1351(2004)019%3C0497:OPTAPS%3E2.0.CO;2

Trewin, N.H. and Welsh, W. 1976. Formation and composition of a graded estuarine shell bed. *Palaeogeography, Palaeoclimatology, Palaeoecology* 19(3): 219–230. https://doi.org/10.1016/0031-0182(76)90015-8

Tucker, B.W. 1944. The ejection of pellets by passerine and other birds. *British Birds* 38: 50–52.

Tulloch, R.J. 1968. Snowy Owls breeding in Shetland in 1967. *British Birds* 61(1): 119–32.

Underhill-Day, J.C. 1985. The food of breeding marsh harriers *Circus aeruginosus* in East Anglia. *Bird Study* 32(3): 199–206. https://doi.org/10.1080/00063658509476880

van Eerden, M.R. and van Rijn, S. 2022. Social Hierarchy within Communal Foraging Flocks of Great Cormorants *Phalacrocorax carbo* as Reflected by Differences in Prey Composition and Food Intake at the Roost. *Ardea* 109(3): 549–563. https://doi.org/10.5253/arde.v109i2.a23

van der Reis, A. and Jeffs, A. 2020. *Determining the diet of New Zealand King Shag using DNA metabarcoding.* Auckland (New Zealand): University of Auckland Report for Department of Conservation.

van Leeuwen, C.H., Lovas-Kiss, Á., Ovegård, M. and Green, A.J. 2017. Great cormorants reveal overlooked secondary dispersal of plants and invertebrates by piscivorous waterbirds. *Biology Letters* 13(10): 20170406.

Village, A. 1990. *The Kestrel.* T. & A.D. Poyser, London.

Votier, S.C., Bearhop, S., Ratcliffe, N. and Furness, R.W. 2001. Pellets as indicators of diet in great skuas *Catharacta skua*. *Bird Study* 48(3): 373–376.

Votier, S.C., Bearhop, S., MacCormick, A., Ratcliffe, N. and Hairness, R.W. 2003. Assessing the diet of great skuas, *Catharacta skua*, using five different techniques. *Polar Biology* 26: 20–26. https://doi.org/10.1007/s00300-002-0446-z

Weiser, E.L. and Powell, A.N. 2011. Evaluating gull diets: a comparison of conventional methods and stable isotope analysis. *Journal of Field Ornithology* 82(3): 297–310. https://doi.org/10.1111/j.1557-9263.2011.00333.x

Whitehead, P.F. 2022. *Gwent Levels Traditional Orchard Invertebrate Study 2019–2021. Report on the invertebrates.* Available on the Living Levels website, https://www.livinglevels.org.uk/orchards-on-the-levels [accessed on 25th April 2023]

Whitfield, D.P., Marquiss, M., Reid, R., Grant, J., Tingay, R. and Evans, R.J. 2013. Breeding season diets of sympatric White-tailed Eagles and Golden Eagles in Scotland: no evidence for competitive effects. *Bird Study* 60(1): 67–76. https://doi.org/10.1080/00063657.2012.742997

Wiersma, W., Piersma, T. and Van Eerden, M.R. 1995. Food intake of great crested grebes *Podiceps cristatus* wintering on cold water as a function of various cost factors. *Ardea* 83: 339–350.

Winkler, A., Nessi, A., Antonioli, D., Laus, M., Santo, N., Parolini, M. and Tremolada, P. 2020. Occurrence of microplastics in pellets from the common kingfisher (*Alcedo atthis*) along the Ticino River, North Italy. *Environmental Science and Pollution Research* 27: 41731–41739. https://doi.org/10.1007/s11356-020-10163-x

Wroot, A.J. 1984. A quantitative method for estimating the amount of earthworm (*Lumbricus terrestris*) in animal diets. *Oikos* 44: 239–242.

Yalden, D. 2009. *The Analysis of Owl Pellets, 4th edition.* The Mammal Society, Southampton.

Yalden, D.W. and Warburton, A.B. 1979. The diet of the Kestrel in the Lake District. *Bird study* 26(3): 163–170.

Young, N.M., Linde-Medina, M., Fondon, J.W., Hallgrímsson, B. and Marcucio, R.S. 2017. Craniofacial diversification in the domestic pigeon and the evolution of the avian skull. *Nature Ecology & Evolution* 1(4): 0095.

LATIN NAMES OF SPECIES

Adder	*Vipera berus*
American mink	*Neogale vison*
Arctic skua	*Stercorarius parasiticus*
Atlantic mackerel	*Scomber scombrus*
Ballan wrasse	*Labrus bergylta*
Bank vole	*Myodes glareolus*
Barn owl	*Tyto alba*
Black rat	*Rattus rattus*
Blackbird	*Turdus merula*
Blue tit	*Cyanistes caeruleus*
Brambling	*Fringilla montifringilla*
Brown hare	*Lepus europaeus*
Brown long-eared bat	*Plecotus auritus*
Brown trout	*Salmo trutta*
Bullhead	*Cottus gobio*
Buzzard	*Buteo buteo*
Carrion crow	*Corvus corone*
Cattle egret	*Bubulcus ibis*
Chaffinch	*Fringilla coelebs*
Chiffchaff	*Phylloscopus collybita*
Chough	*Pyrrhocorax pyrrhocorax*
Chub	*Leuciscus leuciscus*
Cockchafer	*Melolontha melolontha*
Collared dove	*Streptopelia decaocto*
Common clothes moth	*Tineola bisselliella*
Case-bearing clothes moth	*Tineola pellionella*
Common dor beetle	*Geotrupes stercorarius*
Common dumbledor beetle	*Geotrupes spiniger*
Common frog	*Rana temporaria*
Common or brown rat	*Rattus norvegicus*
Common or smooth newt	*Lissotriton vulgaris*
Common or viviparous lizard	*Zootoca vivipara*
Common periwinkle	*Littorina littorea*
Common pipistrelle	*Pipistrellus pipistrellus*
Common shrew	*Sorex araneus*
Common toad	*Bufo bufo*
Cormorant	*Phalacrocorax carbo*
Corn bunting	*Emberiza calandra*
Crowned shrew	*Sorex coronatus*
Curlew	*Numenius arquata*
Cuttlefish	*Sepia officinalis*
Dace	*Leuciscus vulgaris*
Dipper	*Cinclus cinclus*
Dunnock	*Prunella modularis*

Eagle owl	*Bubo bubo*
Edible dormouse	*Glis glis*
European bee-eater	*Merops apiaster*
Feral pigeon	*Columba livia*
Field vole	*Microtus agrestis*
Goldcrest	*Regulus regulus*
Golden eagle	*Aquila chrysaetos*
Goldfinch	*Carduelis carduelis*
Goose barnacles	*Lepas* sp.
Goshawk	*Accipiter gentilis*
Grass snake	*Natrix helvetica*
Great black-backed gull	*Larus marinus*
Great crested newt	*Triturus cristatus*
Great grey shrike	*Lanius excubitor*
Great skua (bonxie)	*Catharacta skua*
Great spotted woodpecker	*Dendrocopos major*
Great tit	*Parus major*
Great white egret	*Ardea alba*
Greater white-toothed shrew	*Crocidura russula*
Green sandpiper	*Tringa ochropus*
Green woodpecker	*Picus viridis*
Greenfinch	*Chloris chloris*
Greenshank	*Tringa nebularia*
Grey heron	*Ardea cinerea*
Grey long-eared bat	*Plecotus austriacus*
Grey squirrel	*Sciurus carolinensis*
Gyrfalcon	*Falco rusticolus*
Haddock	*Melanogrammus aeglefinus*
Harvest mouse	*Micromys minutus*
Hazel dormouse	*Muscardinus avellanarius*
Hedgehog	*Erinaceus europaeus*
Hen harrier	*Circus cyaneus*
Herring	*Clupea harengus*
Herring gull	*Larus argentatus*
Hobby	*Falco subbuteo*
Honey buzzard	*Pernis apivorus*
Hooded crow	*Corvus cornix*
Hoopoe	*Upupa epops*
House martin	*Delichon urbica*
House mouse	*Mus musculus*
House sparrow	*Passer domesticus*
Iberian grey shrike	*Lanius meridionalis*
Jackdaw	*Corvus monedula*
Kestrel	*Falco tinnunculus*
King ragworm	*Allita virens*

Kingfisher	*Alcedo atthis*
Lapland bunting	*Calcarius lapponicus*
Lapwing	*Vanellus vanellus*
Leach's petrel	*Hydrobates leucorhous*
Lesser black-backed gull	*Larus fuscus*
Lesser horseshoe bat	*Rhinolophus hipposideros*
Lesser white-toothed shrew	*Crocidura suaveolens*
Linnet	*Linaria cannabina*
Little egret	*Egretta garzetta*
Little grebe	*Tachybaptus ruficollis*
Little owl	*Athene noctua*
Loggerhead shrike	*Lanius ludovicianus*
Long-eared owl	*Asio otus*
Magpie	*Pica pica*
Marsh harrier	*Circus aeruginosus*
Meadow pipit	*Anthus pratensis*
Merlin	*Falco columbarius*
Minnow	*Phoxinus phoxinus*
Mole	*Talpa europaea*
Mountain hare	*Lepus arcticus*
Mud snail (laver spire shell)	*Peringia ulvae*
Noctule bat	*Nyctalus noctula*
Northern grey shrike	*Lanius borealis*
Orkney, Guernsey or common vole	*Microtus arvalis*
Oystercatcher	*Haematopus ostralegus*
Palmate newt	*Lissotriton helveticus*
Perch	*Perca fluviatilis*
Peregrine	*Falco peregrinus*
Pointed snail	*Cochlicella acuta*
Pygmy shrew	*Sorex minutus*
Rabbit	*Oryctolagus cuniculus*
Raven	*Corvus corax*
Red grouse	*Lagopus lagopus scotica*
Red kite	*Milvus milvus*
Red squirrel	*Sciurus vulgaris*
Red-backed shrike	*Lanius collurio*
Redshank	*Tringa totanus*
Roach	*Leuciscus rutilus*
Robin	*Erithacus rubecula*
Rock ptarmigan	*Lagopus muta*
Roller	*Coracias garrulus*
Rook	*Corvus frugilegus*
Rose chafer	*Cetonia aurata*
Ruffe	*Gymnocephalus cernua*

Sandeel	*Ammodytes marinus*
Sea campion	*Silene maritima*
Sea slater	*Ligia oceanica*
Shag	*Gulosus aristotelis*
Short-eared owl	*Asio flammeus*
Siskin	*Spinus spinus*
Skylark	*Alauda arvensis*
Slow worm	*Anguis fragilis*
Snipe	*Gallinago gallinago*
Snow bunting	*Plectrophenax nivalis*
Snowy owl	*Bubo scandiacus*
Sooty tern	*Onychoprion fuscatus*
Soprano pipistrelle	*Pipistrellus pipistrellus*
Sparrowhawk	*Accipiter nisus*
Squid	*Loligo sp.*
Starling	*Sturnus vulgaris*
Stoat	*Mustela erminea*
Stock dove	*Columba oenas*
Stone loach	*Noemacheilus barbatulus*
Storm petrel	*Hydrobates pelagicus*
Swallow	*Hirundo rustica*
Tawny owl	*Strix aluco*
Three-spined stickleback	*Gasterosteus aculeatus*
Trindade petrel	*Pterodroma arminjoniana*
Turkestan shrike	*Lanius phoenicuroides*
Varied carpet beetle	*Anthrenus verbasci*
Water rail	*Rallus aquaticus*
Water shrew	*Neomys fodiens*
Water vole	*Arvicola amphibius*
Weasel	*Mustela nivalis*
White stork	*Ciconia ciconia*
White-tailed eagle	*Haliaeetus albicilla*
Whiting	*Merlangius merlangus*
Willow ptarmigan	*Lagopus lagopus lagopus*
Wood mouse	*Apodemus sylvaticus*
Woodland dor beetle	*Anoplotrupes stercorosus*
Woodpigeon	*Columba palumbus*
Wren	*Troglodytes troglodytes*
Wryneck	*Jynx torquilla*
Yellowhammer	*Emberiza citrinella*
Yellow-necked mouse	*Apodemus flavicollis*

INDEX

Page numbers in *italics* refer to illustrations and **bold** indicates the main species entry.